PROGRESS IN MATHEMATICS

Volume 10

Mathematical Analysis

PROGRESS IN MATHEMATICS
Translations of Itogi Nauki – Seriya Matematika

1968: Volume 1 – Mathematical Analysis
Volume 2 – Mathematical Analysis

1969: Volume 3 – Probability Theory, Mathematical Statistics, and Theoretical Cybernetics
Volume 4 – Mathematical Analysis
Volume 5 – Algebra

1970: Volume 6 – Topology and Geometry
Volume 7 – Probability Theory, Mathematical Statistics, and Theoretical Cybernetics
Volume 8 – Mathematical Analysis

1971: Volume 9 – Algebra and Geometry
Volume 10 – Mathematical Analysis
Volume 11 – Probability Theory, Mathematical Statistics, and Theoretical Cybernetics

In preparation:
Volume 12 – Algebra and Geometry
Volume 13 – Probability Theory, Mathematical Statistics, and Theoretical Cybernetics
Volume 14 – Algebra, Geometry, and Topology

PROGRESS IN MATHEMATICS

Volume 10

Mathematical Analysis

Edited by

R. V. Gamkrelidze
V. A. Steklov Mathematics Institute
Academy of Sciences of the USSR, Moscow

Translated from Russian by J. S. Wood

SPRINGER SCIENCE+BUSINESS MEDIA, LLC 1971

The original Russian text was published for the All-Union Institute of Scientific and Technical Information in Moscow in 1969 as a volume of *Itogi Nauki – Seriya Matematika*

Library of Congress Catalog Card Number 67-27902
ISBN 978-1-4757-1591-0 ISBN 978-1-4757-1589-7 (eBook)
DOI 10.1007/978-1-4757-1589-7

The present translation is published under an agreement with Mezhdunarodnaya Kniga, the Soviet book export agency

Originally published by Plenum Press, New York in 1971

Preface

The present book contains three articles: "Systems of Linear Differential Equations," by V. P. Palamodov; "Fredholm Operators and Their Generalizations," by S. N. Krachkovskii and A. S. Dikanskii; and "Representations of Groups and Algebras in Spaces with an Indefinite Metric" by M. A. Naimark and R. S. Ismagilov.

In the first article the accent is on those characteristics of systems of differential equations which distinguish the systems from the scalar case. Considerable space is devoted in particular to "nonquadratic systems," a topic that has very recently stimulated interest.

The second article is devoted to the algebraic aspects of the theory of operators (determinant theory in particular) in Banach and linear topological spaces.

The third article reflects the present state of the art in the given area of the theory of representations, which has been receiving considerable attention in connection with its applications in physics (particularly in quantum field theory) and in the theory of differential equations.

Contents

Systems of Linear Differential Equations

V. P. Palamodov

The present article is concerned with research in the last five to ten years on systems of linear partial differential equations. The total number of published works in this area, of course, is too great to cover each one in sufficient detail. While consciously refraining from undertaking such a task, I have endeavored to focus a proportionate amount of attention on each facet of the topic insofar as it embodies the characteristics which distinguish the theory of systems from the analogous theory of one equation in one unknown function. For example, considerable space is accorded the $\bar{\delta}$-Neumann problem, which is endowed with a specialized character and whose solution has contributed a great deal that is conceptually new to the general theory. On the other hand, the highly-developed theory of boundary-value problems is scarcely touched at all, as its methods pertain by and large to scalar theory.

§ 1. Quadratic Systems of Equations

We interpret the following as a quadratic system:

$$\sum_{1}^{N} p_{ij}(x,\ D)\, u_j(x) = f_i(x), \quad i = 1, \ldots, N, \tag{1.1}$$

in which the number of equations N is equal to the number of unknown functions and the determinant of the characteristic matrix $p(\chi, \xi) = \{p_{ij}(\chi, \xi)\}$ is not identically equal to zero. Of all the systems of differential equations, the class of quadratic systems is most nearly akin to scalar systems (N = 1) in its properties and has been the most fully investigated. The study of quadratic systems, for example the local properties of solutions and the solvability of boundary-value problems, is approached, as in the scalar case, on the basis of the delineation of elliptic, hyperbolic, and other types. The type-discrimination process, in turn, is based on the separation of the principal part of the operator $p = \{p_{ij}\}$, but this, as opposed to the scalar case, is a substantive problem.

The most general method for separating the principal part of the operator p corresponding to a quadratic system has been proposed by Douglis and Nirenberg [73]. Their method entails the following: Let integers $s_1, \ldots, s_N$ and $t_1, \ldots, t_N$ be chosen such that $\deg p_{ij} \leq s_i + t_j$ for any i and j. We denote by $\bar{p}_{ij}$ the sum of terms of order $s_i + t_j$ in the operator p_{ij}; the matrix $\bar{p} = \{\bar{p}_{ij}\}$ is called the principal part of p. It is clear that $\bar{p}$ depends on the choice of numbers s_i and t_j, and the latter do not contain useful information on the operator p for every choice. Douglis and Nireberg have proved that if for a certain choice of s_i and t_j the determinant of $\bar{p}(x, \xi)$ is the elliptic operator symbol, then system (1.1) is subject to the classical theorems on the regularity of elliptic equations. Specifically, there is adequate correspondence between the smoothness of the right-hand sides of f_i with regard for their weights s_i and the smootheness of the solutions u_j with regard for their weights t_j. In [73] a suitable a priori estimate is also established. For the same class of systems Morrey and Nirenberg [122] have proved the analyticity of the solutions on the assumption of analyticity of the right-hand sides and coefficients of the operators p_{ij}.

Volevich [8], in an analysis of the Douglis-Nirenberg concept of the principal part of a system, has shown that the numbers s_i and t_j can always be chosen so that the determinant of the matrix $\bar{p}(x, \xi)$ proves to be equal to the sum of the highest-order terms that occur under the standard technique for computing the determinant of the matrix $p(x, \xi)$. Consequently, under Volevich's rule for the selection of the weights the matrix $\bar{p}(x, \xi)$ is degenerate if and only if the highest terms in the expression for $\det p(x, \xi)$ cancel one another. Related to this type of degeneracy is the example given by Volevich in [9] of a system with $N = 2$ and two independent variables. In this example $\det p(x, \xi)$ is the elliptic operator symbol, but the homogeneous system $pu = 0$ has solutions of finite smoothness.

In [11] Volevich delineates a class of hypoelliptic systems, which includes systems of pseudodifferential equations as well and is analogous to the Douglis-Nirenberg class of elliptic systems. The most important of the conditions imposed on the operator symbol for the proof of the regularity theorem in [11] is the fact that there exist weight functions $\mu_i(\xi)$ and $\nu_j(\xi)$, i, j = 1,..., N subject to specific conditions such that for any i and j the following inequalities hold:

$$|p_{ij}(x, \xi)| \leqslant \mu_i(\xi)/\nu_j(\xi)$$

and

$$|\det p(x, \xi)| \geqslant c \prod_1^N \mu_i(\xi)/\nu_j(\xi).$$

Partially-hypoelliptic systems have also been studied in [10].

Boundary-value problems for Douglis-Nirenberg elliptic systems have been investigated by Volevich [10], Solonnikov [46, 47], and Roitberg and Sheftel' [43]. These authors consider boundary-value problems satisfying the generalized Shapiro-Loptinskii condition; boundary-value problems of this kind are also called regular or coercive problems. Earlier Solomyak [45] formulated an example of a first-order elliptic system with four independent variables for which there is no regular boundary-value problem. Atiyah and Bott [60] have found a necessary and sufficient condition in order for a given elliptic operator to have a regular boundary-value problem on a given manifold with boundary. Vishik and Eskin [7] have developed a theory of boundary-value problems for elliptic systems of pseudodifferential equations.

In connection with the investigation of irregular boundary-value problems Vainberg and Grushin [5] have analyzed a class of quadratic systems of pseudodifferential equations which is broader than the class of elliptic systems. In [5] the pseudodifferential operator $p = \{p_{ij}\}$ acts in the space of cross sections of an N-dimensional trivial fiber bundle over a compact differential manifold X; every scalar operator p_{ij}, correct to an arbitrarily low-order operator, is written by means of the Fourier transform in terms of its complete symbol $p_{ij}(x, \xi)$, which in general represents an infinite formal sum of functions, smooth for $\xi \neq 0$ and homogeneous on ξ, which are defined on the cotangent bundle $T^*(X)$. The highest order of the homogeneous terms in the complete symbols of the operators p_{ij} is assumed to be finite and is called the order of p. We recall that the standard definition of the elliptic pseudodifferential operator states that the matrix $\bar{p}$ formed by the homogeneous terms of order $m = \deg p$ of the complete symbols $p_{ij}(x, \xi)$ is assumed to be nondegenerate at every point $(x, \xi) \in T^*(X)$, where $\xi \neq 0$. The class of operators designated by the authors as uniformly nonelliptic is defined by the following weaker condition: For any point $x \in X$ and any point $\xi \in T_x^*(X)$, $\xi \neq 0$, there exist in the neighborhood of x elliptic pseudodifferential operators a and b, acting in the same bundle as p, such that the complete symbol of the operator apb in the neighborhood of the point $(x, \xi) \in T^*(X)$ has a partitioned-diagonal form, where all partitions are symbols of certain elliptic operators whose orders assume all values from m to $m - k$. As shown by the authors, the number k, as well as the dimensions of the partitions are independent of the point (x, ξ) if $\dim X > 1$ and X is connected.

Vainberg and Grushin have found an equivalent condition more suitable for verification. It calls for the existence of a pseudodifferential rectangular matrix d on X of special type such that the composition dp is an elliptic operator. With the aid of this matrix the authors introduce a special family H_p^s, $-\infty < s < \infty$, of functional spaces on X such that for any s the space H_p^s is contained between the Sobolev classes H^{s-m} and H^{s-m+k} and the operator p acts continuously from H^s into H_p^s. The fundamental result of Vainberg and Grushin in [5] is the fact that this operator is Noetherian for any s, i.e., its kernel and cokernel are finite-dimensional. In [6] Vainberg and Grushin investigate a certain class of irregular boundary-value problems for elliptic systems, reducing them to the investigation of elliptic pseudodifferential operators of the type described, at the boundary.

Irregular boundary-value problems for elliptic systems have also been studied by Tovmasyan [51, 52].

Dezin [19] has pointed out that a quadratic system of differential equations cannot have a correct boundary-value problem in the sense of Vishik, even when the coefficients of the operator are constant. The existence of such a problem in a given domain Ω is equivalent to the fulfillment of the following two inequalities in L_2-norms:

$$\|u\| \leqslant c\|pu\|, \quad \|u\| \leqslant c\|p^*u\|$$

on the set of infinitely-differentiable functions u whose supports belong to Ω. Independently of one another Paneyakh [41] and Fuglede

[79] found a necessary and sufficient condition on a operator p with constant coefficients in order for the above inequalities to hold. The condition states that every minor q of order N−1 of the characteristic matrix p (ξ) must satisfy the inequality

$$|q(\xi)|^2 \leqslant c \sum_{\alpha} |d^{(\alpha)}(\xi)|^2, \quad d = \det p.$$

Both authors also find a condition for a given matrix operator r to be subordinate to the operator p.

We conclude with a summary of recent papers on aspects of the problem not covered in this section. The solvability and properties of the solutions of the Cauchy problem for quadratic systems have been investigated by Gorin [18], Fridlender [53], Foias and Gussi [77], Kopaček and Suha [26], Kopaček [101], Sochneva [48], Schaefer [127], Yamamoto [140], and Yamanaka [141]. Initial-value problems of the Goursat-Riquier type for quadratic systems after Ledněv [28] have been studied by Persson [125] and Garding [82]. Similar problems have also been investigated by Vaillant [139], Kovach [25], and Tricomi [138].

Boundary-value problems for positive symmetric and allied first-order systems, as well as boundary-value problems for hyperbolic systems have been investigated by Phillips and Sarason [126], Friedrichs and Lax [78], Galakhov [13], Brenner [63], and Arganovich [2, 3].

§ 2. Systems with Constant Coefficients

The main preoccupation in this area has been with the solvability problem for the general inhomogeneous system

$$p(D)u = f, \tag{2.1}$$

where p is a rectangular matrix of arbitrary dimensions formed by differential operators with constant coefficients, as well as the problem of the exponential representation of the solutions of the corresponding homogeneous system

$$p(D)u = 0. \tag{2.2}$$

The compatibility conditions necessary for the solvability of (2.1) are written as follows. Let A be the ring of all differential operators with constant coefficients, i.e., the ring of polynomials over C of the operators $\partial/\partial x_1, \ldots, \partial/\partial x_n$. For every positive integer k we denote by A^k the A-module equal to the direct sum of k copies of the indicated ring. Then we consider the mapping of A-modules p': $A^t \to A^s$, the action of which consists in the right multiplication of the vectors from A^t by the matrix p (t is the number of rows, s the number of columns of p). The Noetherian character of A implies the existence of a matrix q formed by elements of A such that the map q': $A^r \to A^t$ coincides with the kernel of p'. The sought-after compatibility condition is written in the form of a homogeneous system of equations with constant coefficients:

$$q(D)f = 0. \tag{2.3}$$

It follows from the construction of the operator q that qp = 0, so that the fulfillment of (2.3) is necessary for the solvability of (2.1) and, in addition, system (2.3) contains all the linear relations with coefficients from A between the right-hand sides of (2.1) which are implied bythe relations between rows of the matrix p.

Let Φ be a functional space in which the action of all differential operators with constant coefficients is defined. System (2.1) is said to be solvable in this space if for any right-hand side $f \in \Phi^t$ satisfying the compatibility system (2.3) there exists a solution $u \in \Phi^s$. In the case when (2.1) is scalar, i.e., when s = t = 1, or when it is quadratic, i.e., when s = t and $\det p(\xi) \not\equiv 0$, the corresponding compatibility condition is empty. In this relatively simpler case the solvability problem was investigated back in 1952 and 1953 by Ehrenpreis and Malgrange. They verified the solvability of such systems in the class $\mathbf{E}(\Omega)$ of all infinitely-differentiable functions and in the class $\mathbf{D}'(\Omega)$ of all generalized functions defined in an arbitrary convex domain Ω.*

Prior to the Sixties the solvability of nonquadratic systems had only been studied for certain first-order systems: systems associated with the exterior differentiation operator (inverse theorem of Poincaré); systems associated with the operator $\bar{\partial}$ (Grothendieck lemma, theorem in the theory of coherent sheaves); combinations of such systems (Nirenberg [124]). The first general result was ob-

*This result was first verified for the space $\mathbf{D}'(\xi)$ by Ehrenpreis in the case $\Omega = R^n$ and was proved for an arbitrary convex domain Ω by Malgrange in 1958-59. A bibliography pertaining to the scalar theory is contained in [37].

tained in 1959-60 by Malgrange [106-109], who verified the solvability of any system in the class S' of generalized slow-growth functions in $\mathbf{R}^n$ (see also [35]). For the classes of rapid-growth functions the solvability of system (2.1) was established in 1962-63 independently by Malgrange [110] [spaces $\mathbf{E}(\Omega)$ and $\mathbf{D}'(\Gamma)$, where Γ is any convex compactum] and Palamodov [36] [spaces $\mathbf{E}(\Omega)$ and $\mathbf{D}'(\Omega)$, where Ω is any convex domain].

As noted by Malgrange [111], if the domain Ω is connected, its convexity is also a necessary condition in order for any system with constant coefficients to be solvable in class $\mathbf{E}(\Omega)$.

In [112] Malgrange investigates the solvability problem in the case when the operator p is specified in an even-dimensional space equipped with the structure of complex Euclidean space and contains only "antiholomorphic" differentiations. He shows that for any such operator system (2.1) is solvable in $\mathbf{E}(\Omega)$ for any holomorphically-convex domain Ω. The class of holomorphically-convex domains is the broadest class in which such a statement can hold true, because, according to the classical theorems of the theory of analytical functions of several variables, the solvability of systems of differential equations associated with the operator $\bar{\partial}$ in $\mathbf{E}(\Omega)$ requires as a necessary (and sufficient) condition that the domain be holomorphically convex. Palamodov in [37] translates this theorem of Malgrange to the space $\mathbf{D}'(\Omega)$.

A number of results in the solvability problem have been obtained by Palamodov in [37], in which a particular version of the problem called the M-convexity problem is investigated. This approach is based on the language of homological algebra, which was first set forth in the indicated context by Malgrange [111]. Associated with system (2.1) is an A-module M of finite type, equal to the cokernel of the mapping described above, $p' \colon A^t \to A^s$; the solvability of (2.1) in the functional class Φ is equivalent to the equation $\mathrm{Ext}^1_A(M, \Phi) = 0$. This notation implies that the solvability of system (2.1) in the given class Φ depends only on M and not on the matrix p itself. In general, all significant properties of (2.1) are retained under replacement of the operator p by any other operator p_1 such that $\mathrm{Coker}\, p_1' \cong M = \mathrm{Coker}\, p$. The same is true of system (2.2), because the space Φ_p formed by the solutions of (2.2) in class Φ^s is isomorphic as an A-module of the space $\mathrm{Hom}_A(M, \Phi)$. We note also that the characteristic set $N \subset \mathbf{C}^n$ of the operator p is also a function of M; it coincides with the support of that module.

We now formulate the M-convexity problem per se. We choose an arbitrary A-module M of finite type and its free resolvent, i.e., an exact sequence of the form

$$M_*: \ldots \to A^r \xrightarrow{q'} A^t \xrightarrow{p'} A^s \to M \to 0 \qquad (2.4)$$

[the matrix p defines a system of the form (2.1) with which is associated the module M; the matrix q determines the compatibility system for (2.1), etc.]. The functional space Φ is said to be M-convex if it is an A-module and $\mathrm{Ext}_A^i(M, \Phi) = 0$ for all $i \geq 1$. An additional requirement is that all mappings in the complex $\mathrm{Hom}_A(M_*, \Phi)$ be topological homomorphisms; if Φ is a Frechet space, this property is implicit in the first condition. The space Φ is said to be strongly M-convex if it is M-convex and, in addition, the exponential polynomials satisfying system (2.2) form a dense subset in the space Φ_p. The latter, according to the above arguments, depends on the modules M and Φ, but not on the choice of the matrix p in (2.4). The center of attention in [37] is the M-convexity problem for classes $\mathbf{E}(\Omega)$ and $\mathbf{D}'(\Omega)$. The above-formulated general result of Malgrange and Palamodov is stated as follows in precise form: The indicated classes are strongly M-convex for any convex domain Ω and module M of finite type. It follows from this assertion that for any domain $\Omega \subset \mathbf{R}^n$ the isomorphism

$$\mathrm{Ext}_A^i(M, \mathbf{D}'(\Omega) \cong H^i(\Omega, \theta), \quad i=0,1,2,\ldots)$$

holds, where θ is the sheaf in Ω of germs of generalized solutions of (2.1) (generalization of the Dolbeault isomorphism). Proceeding from this isomorphism and invoking the reasoning of Serre in his study of the topology of Runge domains, Palamodov found a necessary condition on the domain Ω for M-convexity and strong M-convexity of $\mathbf{D}'(\Omega)$. This condition states that the Betti number of Ω must be equal to zero in dimensions greater than or less than, respectively, the number $\dim_C N$. The same assertions apply to $\mathbf{E}(\Omega)$.

The sufficient conditions deduced in [37] for the M-convexity of $\mathbf{D}'(\Omega)$ are remote from the necessary condition stated above. One of those sufficient conditions is formulated as follows: The domain Ω admits a locally-finite covering formed by convex domains intersecting at most on the $(K + 1)$th, where $0 < k \leq n$, and the module M can be embedded in an exact sequence of A-maps of the form

$$0 \to M \to A^{s_1} \to A^{s_2} \to \ldots \to A^{s_k}.$$

Another result bearing on the M-convexity problem pertains to k-convex domains. The k-convexity condition on Ω, being analogous to the k-pseudoconvexity condition, states that there exists in Ω a real function φ of class C^2 for which the matrix of second derivatives has at least k positive eigenvalues at every point $x \in \Omega/K$, where $K \subset \Omega$ is a compactum such that the set of lesser values $\{x \in \Omega, \varphi(x) \leq c\}$ is a compactum for any c. The following theorem has been proved in [37]: If the domain Ω is k-complex and the module M is hypoelliptic, i.e., if the mapping $\zeta \to \operatorname{Im} \zeta$, specified on $N \subset C^n$, is proper, then all spaces $\operatorname{Ext}_A^i(m, D'(\Omega))$, $i > n - k$, are finite-dimensional. But if the compactum K involved in the k-convexity condition on Ω is empty, all those spaces are trivial. This result is the analog of the Andreotti-Grauert finiteness theorem [56].

The result obtained by Malgrange in [113] is also related to the theorems of Andreotti and Grauert [56]. The problem investigated by Malgrange in [113] is somewhat more general than the M-convexity problem; he considers an arbitrary complex of A-maps of the form

$$M_*: \ldots \to A^{t_i} \to A^{t_{i+1}} \to \ldots,$$

in which only a finite number of moduli are nonvanishing, and analyzes the cohomologies of the complex $H_A(M*, \mathbf{E}(\Omega))$. We note that if $M*$ is the free resolvent of some module M, then the indicated cohomologies coincide with the modules $\operatorname{Ext}_A^i(M, \mathbf{E}(\Omega))$, $i = 0, 1, 2, \ldots$. Malgrange formulates the condition on Ω in terms of certain cohomologies associated with the complex $M*$ in the space of infinitely-differentiable functions whose supports are contained in an arbitrarily small neighborhood of the boundary of Ω. The fundamental result of [113] is that under fulfillment of the foregoing condition and the assumption that all modules $H_i(M*)$ are hypoelliptic the spaces $H^i(\operatorname{Hom}_A(M*, \mathbf{E}(\Omega))$ are finite-dimensional for definite values of i. We note that the results of Andreotti and Grauert, like analogous results of Malgrange and Palamodov, are based on a notion advanced by Enrenpreis [74].

Kamatsu [100] has investigated the solvability of system (2.1) in a class of hyperfunctions. If Ω is any domain in $\mathbf{R}^n$ and $n > 1$, then the space of hyperfunctions in Ω, according to Sato, is isomorphic to the $(n-1)$-dimensional cohomology of the sheaf of germs of holomorphic function in ω/Ω, where ω is any holomorphicity domain in the enveloping complex space such that $\omega \cap R^n = \Omega$. Every generalized function of finite or infinite order in Ω is also a hyperfunction. An important property of the sheaf of germs of hyperfunctions is the fact that it is slack; this is not an intrinsic property of the sheaf of germs of generalized functions. Komatsu has shown that any system (2.1) is solvable in the space of hyperfunctions defined in an arbitrary convex domain; this assertion is a consequence of the analogous result for class $\mathbf{E}(\Omega)$. The foregoing brought Komatsu to the conclusion that for any coherent sheaf on a complex analytic manifold of dimension n there exists a slack resolvent of length n.

The M-convexity problem and several other aspects of the theory of systems of differential equations with constant coefficients are presented in [37] as a series of corollaries to the central theorem, supporting the feasibility of the exponential representation of all solutions of the homogeneous system (2.2) for any operator p. The exponential representation, in turn, is a consequence of a particular theorem concerned with polynomial submodules in spaces of integral vector functions subject to special growth constraints at infinity. An approach to the theory based on this type of theorem was first proposed by Ehrenpreis in his "Fundamental Principle" published in 1960 [75]. The exponential representation theorem was announced by Palamodov in [36] and subsequently proved in [37] and [39] for classes of rapid growth functions. We note the presentation of the exponential representation theorem in implicit form for class S' in [35].

We now state the result of [39], which is more precise than that of [37]. The following definitions are needed. We interpret A as a ring of polynomials with complex coefficients in Euclidean space $\mathbf{C}^n$, in which the coordinates are denoted by the symbol $\varsigma = (\varsigma_1, \ldots, \varsigma_n)$. Let S be a primary submodule of the A-module A^k, and let I be its radical.* Every $k \times l$ matrix $d(\varsigma, \partial/\partial\varsigma)$ formed by the

*Recall that the radical of a submodule $S \subset T$ is defined as the ideal $I \subset A$ formed by all ring elements whose action in T/S is nilpotent. The submodule S is said to be primary if it follows from $at \in S$, where $a \in A$ and $t \in T$, that either $a \in I$ or $t \in S$.

polynomials of c and the differential operators $\partial/\partial\varsigma_1, \ldots, \partial/\partial\varsigma_n$, can be set in correspondence with a differential operator d: $A^k \to A^l$. Let us examine the sequence of linear maps

$$0 \to S \to A^k \overset{\pi d}{\to} [A/I]^l,$$

in which π: $A^l \to [A/I]^l$ is a canonical mapping. We say that the differential operator d is associated with the submodule S if the completion of the above sequence on the m-type topology is an exact sequence of linear spaces for any maximal ideal $M \subset A$. Every primary submodule S has associated with it at least one operator, and all such operators are in a definite sense mutually equivalent.

Consider the mapping of A-modules p': $A^t \to A^s$ associated with system (2.1). By the foregoing here $p = p(\zeta)$ is the characteristic matrix of the operator p (D). Let $P_1 \cap \ldots \cap P_\Lambda$ be the reduced primary representation of the submodule $p'A^t \subset A^s$,* for every $\lambda = 1, \ldots, \Lambda$ let N_λ be an algebraic subset in $\mathbf{C}^n$ of roots of the radical of the submodule P_λ, and let d_λ: $A^s \to A^{s_\lambda}$ be the differential operator associated with that submodule. The following theorem is deduced in [39]: If Ω is a convex domain in R^n, then every generalized function u in Ω that satisfies (2.2) in Ω can be written in the form

$$u(x) = \sum_{\lambda=1}^{\Lambda} \int d_\lambda(\zeta, \partial/\partial\zeta) \exp(\zeta, -ix)\, \mu_\lambda(\zeta), \tag{2.5}$$

where for every λ μ_λ is complex additive measure concentrated on N_λ, such that the integrals in (2.5) are absolutely convergent in the space $[D'(\Omega)]^s$. The latter condition implies that for any compactum $K \subset \Omega$ there exists a constant q such that all integrals

$$\int (|\zeta| + 1)^{-q} \exp(\gamma_K(\operatorname{Im}\zeta))\, |\mu_\lambda(\zeta)|, \quad \lambda = 1, \ldots, \Lambda, \tag{2.6}$$

are finite; γ_K is the support function of K. If u is an infinitely-differentiable function, the measures μ_λ can be chosen so as to render the integrals (2.6) finite for all K and q. This representation is adequate in the sense that for any measures with finite integrals (2.6) the right-hand side of (2.5) is a generalized and, accordingly, infinitely-differentiable solution of system (2.2) in Ω. In order to

*This means that $p'A^t = P_1 \cap \ldots \cap P_\Lambda$, all P_λ are primary submodules of A^s, and none of them contains intersections of the others.

deduce the solvability of (2.1) in any convex domain from this theorem it is sufficient to apply the theorem to the homogeneous system (2.3) and its associated operator p': $A^t \to A^s$. In [37] an analogous, though relatively less precise, representation is established for classes of solutions in $\mathbf{R}^n$ with definite growth restrictions at infinity and for the generalizations of solutions of infinity and for the generalizations of solutions of infinite order. The method [37] permits the same thing to be done for a number of other functional spaces.

In [37] the exponential representation is a universal method for the investigation of the local properties of the solutions of (2.2). The substance of the method lies in the fact that the elementary properties of the Fourier transform enable one to connect the smoothness of a solution u with the decomposition in $\mathbf{C}^n$ of the supports of the measures $\mu\lambda$ in its representation (2.5) and, hence, also with the decomposition of the algebraic set $N = \cup N_\lambda$ comprising the characteristic set of the operator p. Analogous reasoning is used to prove the uniqueness of the solution of the generalized Cauchy problem for (2.2) with multidimensional time and initial data on the characteristic subset. Essentially by virtue of the a priori growth limitations imposed at infinity the solution of (2.2) turns out in a certain sense to be analytically dependent on the time, whence follows its unique dependence on the initial data. We note that the uniqueness theorem deduced in [37] is a generalization of the well-known result of Gel'fand and Shilov [14].

Also investigated in [37] are the characteristic properties of the solutions of (2.2) with respect to overdetermined operators p. Among those properties are the possibility of extending the solution into a larger domain, the unique and continuous dependence of the solution on its values in the neighborhood of a small portion of the boundary of the domain of definition, and the extension of smoothness from that neighborhood onto the entire domain.

The simplest problem involving the extension of solutions is formulated as follows: Let there be given a domain Ω and a compactum $K \subset \Omega$; it is required to characterize the set of solutions of system (2.2) in $\Omega \setminus K$ that can be extended to Ω as solutions of the same system. This problem is solved on the assumption that K is convex in the sense that an isomorphism is established between the barrier to that extension and the module equal to the tensor product over A of the module Ext_A^1 (M, A) and the space of integral functions in $\mathbf{C}^n$ with definite growth restrictions at infinity. It follows from this result, in particular, that in order for every solution defined in $\Omega \setminus K$ to be uniquely extended as a solution in Ω it is necessary and sufficient that $\mathrm{Hom}_A(M, A) = \mathrm{Ext}_A^1(M, A) = 0$. The latter relations are equivalent to the inequality $\dim_{\mathbf{C}} N < n-1$. The operators p satisfying these relations are said in [37] to be overdetermined. The above-indicated isomorphism is also used to find the conditions on the operator such that every one-point singularity of the smooth solutions of (2.2) will be removable. The scalar operators endowed with this property have been described by V. V. Grushin.

We arrive at the following from other results concerning the possibility of extension. Let Ω be a convex domain, let $K \subset \Omega$ be a compactum, and let Σ_i, $i = 1,\dots, k$, where $0 < k < n$, be closed half-spaces in $\mathbf{R}^n$. Let us assume that the conditions $\mathrm{Ext}_A^i(M, A) = 0$, $i = 1,\dots, k$, are fulfilled. Then for every solution u of (2.2) defined in $\Omega \setminus (K \cup \Sigma)$, where $\Sigma = \cup \Sigma_i$, it is possible to find a solution of the same system, defined in Ω Σ and coinciding with u in the domain $\Omega \setminus (K' \cup \Sigma)$, where $K' \subset \Omega$ is a compactum. In the case $k = 1$ this result also appears in Malgrange's paper [111].

A theorem on the extension of infinite differentiability is stated in the same notation: If all modules $\mathrm{Ext}_A^i(M, A)$, $i = 0,\dots, k$, are hypoelliptic, then every generalized solution of (2.2) defined in $\Omega \setminus \Sigma$ and infinitely differentiable in $\Omega \setminus (K \cup \Sigma)$ is infinitely differentiable in the entire domain $\Omega \setminus \Sigma$. We cite an earlier result of Zerner [142], which sets forth the limit of possible amplifications of the foregoing theorem. Zerner introduces the concept of the simple bicharacteristic of an operator p; we formulate its definition, assuming for simplicity that the matrix of p contains one column, i.e., $s = 1$ and $p = (p_1,\dots, p_t)$. If ξ is a real point of the characteristic set N, then the bicharacteristic $V(\xi)$ is defined as the linear space spanned by the t vectors $\mathrm{grad}\,\bar{p}_j(\xi)$, $j = 1,\dots, t$, where $\bar{p}_j$ is the principal part of the polynomial p_j [the coefficients of the polynomials p_j are considered to be real]. The bicharacteristic $V(\xi)$ is said to be simple if its dimension is equal to t. Zerner's result is stated as follows: If V is a simple bicharacteristic of the operator p, then system (2.2) has a solution in $\mathbf{R}^n$ that is infinitely differentiable in $\mathbf{R}^n \setminus V$ but is not infinitely differentiable in $\mathbf{R}^n$.

§ 3. Spencer Constructions and the Local Solvability of Systems with Variable Coefficients

The recent literature dealing with nonquadratic systems of equations with variable coefficients is tied in one way or another with the problem of the solvability of the system

$$p(x, D)u=f. \tag{3.1}$$

For such systems, as opposed to the case of constant coefficients, the search for a complete set of compatibility conditions in the form of a system of differential equations

$$q(x, D)f=0 \tag{3.2}$$

is now a substantive problem. One version of the solvability problem is formulated as follows. Consider the differentiable manifold X, the vector bundle E and F over X and the differential operator p: $\widetilde{E} \to \widetilde{F}$.* In this case q is a differential operator acting from $\widetilde{F}$ into $\widetilde{G}$, where G is also a vector bundle over X such that qp = 0 (the operator q will be elaborated upon presently). The local solvability problem refers to the investigation of the complex of sheaves

$$\widetilde{E}\xrightarrow{p}\widetilde{F}\xrightarrow{q}\widetilde{G}, \tag{3.3}$$

more particularly to the search for conditions guaranteeing its accuracy. The global solvability problem entails the investigation of the homology of the complex

$$\Gamma(E)\xrightarrow{p}\Gamma(F)\xrightarrow{q}\Gamma(G),$$

where, for example, Γ(E) is the space of cross sections of a particular type of the bundle E.

This problem area has been represented in different versions in recent years primarily by the work of American mathematicians along lines similar to Spencer's: Quillen, Sternberg, Guillemin, Sweeney, Goldschmidt, Mansfield, Cenkl, and MacKichan. Typical of their work is the application of the constructions proposed by Spencer involving the sheaf Θ of germs of solutions of the system of equations pu = 0, in particular the so-called Spencer resolvents of that sheaf [129, 130].

*We denote by $\widetilde{E}$ the sheaf of germs of cross sections of the vector bundle E.

The construction of the first Spencer resolvent comprises the following. Let $J^k(E)$ be the bundle of jets of order k (k-jets) of the bundle E, and let $j^k: \tilde{E} \to J^k(E)$ be a canonical differential operator.* If the order of the given differential operator p is m, then the bundle map $p_m: J^m(E) \to F$ such that $p = \tilde{p}_m j^m$ is uniquely determined, where $\tilde{p}_m: J^m(E) \to F$ is the mapping, associated with p_m, of the sheaves of germs of cross sections. The differential operator of order m + k equal to the composition of p and the canonical operator $j^k: \tilde{F} \to \overline{J^k}(F)$ is called the kth extension of p. Let $p_{m+k}: J^{m+k}(E) \to J^k(E)$ be the bundle corresponding to the operator $j^k p$, and let $R_{m+k} = \operatorname{Ker} p_{m+k}$.

For every $k \geq 0$ there is defined a bundle map $\pi_k: J^k(E) \to J^{k-1}(E)$ referring the jet of order k to the jet formed by its terms of order less than k. This mapping takes R_k into R_{k-1}. According to Spencer [132] a differential operator p is said to be regular if there exists an integer k_0 such that R_{k_0} is a vector subbundle of the bundle $J^{k_0}(E)$ and the mapping $\pi_k: R_k \to R_{k-1}$ is epimorphic for any $k > k_0$. It follows from the regularity of p that R_k and $g_k = \operatorname{Ker}\{R_k \to R_{k-1}\}$ are vector bundles for any $k > k_0$. This definition has fundamental significance in the theory of Spencer; all the results formulated below pertain exclusively to regular operators.

For every pair (k, i) we consider the bundle R_k^i, which is equal to the tensor product over C of the bundle R_k and the bundle of exterior differential forms on X of order i; the bundle g_k^i is constructed analogously. The first ("naive") resolvent of Spencer is the complex of sheaves†

$$0 \to \theta \xrightarrow{j^k} \tilde{R}_{k+n} \xrightarrow{D} \tilde{R}^1_{k+n-1} \xrightarrow{D} \dots \to \tilde{R}^n_k \to 0, \tag{3.4}$$

in which D is a certain first-order differential operator. The opera-

*We recall the definition of the bundle $J^k(E)$. If $x \in X$ and U_x is a sufficiently small neighborhood of x, then the bundle $J^k(E)$ over U_x is isomorphic to the right product of U_x and the space of segments of the Taylor series to order k at x of the elements $f \in \tilde{E}_x$. The automorphism of a fiber of $J^k(E)$ under the transition from U_x to U_y is determined by the automorphism of a fiber of E and the rule for transformation of the coefficients of the Taylor expansion of a scalar function on X due to the corresponding change of coordinates. The map j^k sets the germ $f \in \tilde{E}_x$ in correspondence with the segments of its Taylor series at points near x.

†Note that this complex, in general, is not a resolvent per se, i.e., it is not an exact sequence (see below).

tor D takes the subsheaf $\widetilde{g}^i_{k+1}$ into $\widetilde{g}^{i+1}_k$ and acts accordingly as a zero-order differential operator. Corresponding to this zero-order differential operator is a bundle map $g^i_{k+1} \to g^{i+1}_k$, which is denoted by δ. Consequently, we arrive at the complex of vector bundles

$$0 \to g_{k+n} \xrightarrow{\delta} g^1_{k+n-1} \xrightarrow{\delta} \dots \xrightarrow{\delta} g^n_k \to 0, \tag{3.5}$$

which is called the Spencer δ-sequence. Spencer in a special case, and later Quillen* in the general case, proved that the δ-sequence is exact for $k > k_1$, where k_1 is a constant depending only on the dimension of X, E, and the order of p. Sternberg noted a connection between the δ-sequence and the involutiveness concept used by Kartan in his theory of the integration of Pfaff systems, and he incorporated this concept into Spencer's theory. Serre has proved that the involutive character of an operator is equivalent to exactness of the δ-sequence (Serre's arguments are presented in [85]; analogous arguments may be found in [137]).

The following important result of Quillen stems from the investigation of the δ-sequence: For every regular operator p there exists a regular operator q: $\widetilde{F} \to \widetilde{G}$ such that sequence (3.3) is formally exact, i.e., for any $k \geq l = \deg q$ the sequence $J^{k+m}(E) \xrightarrow{p_{k+m}} J^k(F) \xrightarrow{q_k} J^{k-l}(G)$, obtained from (3.3) by the extension operation described above is exact. This assertion implies that for any point $x \in X$ and elements $f \in J^k_x(F)$, i.e., any segment of the Taylor series at x of some cross section of the bundle F, the condition $q(x+\xi, D_\xi)f(\xi) = 0$ is necessary and sufficient for the solvability of the system of equations $p(x+\xi, D_\xi)u(\xi) = f(\xi)$, $u \in J^{k+m}_x(E)$. Thus constructed, the differential operator q realizes the formal compatibility condition (3.2) discussed in the beginning of the present section. This means that for any point $x \in X$ and vector $r(x, D)$ formed by differential operators with smooth coefficients and defined in a neighborhood of x such that $rp = 0$ the relation $rf = 0$ is a consequence of system (3.2). Quillen has also shown that the exactness of the first Spencer resolvent in the term $\widetilde{R}^1_{n+k-1}$ for $k > k_1$ is equivalent to the exactness of sequence (3.3), i.e., the local solvability of system (3.1). Similarly, the exactness of sequence (3.4) in the term $\widetilde{R}^2_{n+k-2}$ is equivalent to the solvability of the

*In all references to D. G. Quillen I have in mind his unpublished dissertation: "Formal Properties of Overdetermined Systems of Linear Partial Differential Equations," Harvard (1963).

inhomogeneous system $qf = g$ under the compatibility condition $q_1 g = 0$ corresponding to that system, etc.

The second ("sophisticated") Spencer resolvent is construeted as follows. As we observed, the δ-sequence (3.5) is exact for all sufficiently large k. Therefore, the kernel of the mapping δ: $g^i_{k+1} \to g^{i+1}_k$ in that sequence is a vector subbundle of the bundle g^i_{k+1} and, hence, of the bundle R^i_{k+1}. The factor bundle R^i_{k+1} on that subbundle is denoted by C^i_k; it is readily verified that the differential operator D: $\widetilde{R}^i_{k+2} \to \widetilde{R}^{i+1}_{k+1}$ of the first resolvent (3.4) generates a first-order differential operator D: $\widetilde{C}^i_k \to \widetilde{C}^{i+1}_k$ such that $DD = 0$. The resulting complex of sheaves

$$0 \to \theta \to \widetilde{C}^0_k \xrightarrow{D} \widetilde{C}^1_k \xrightarrow{D} \ldots \to \widetilde{C}^n_k \to 0, \quad C^0_k = R_{k+2} \qquad (3.6)$$

is called the second Spencer resolvent. The motivation behind the introduction of this sequence of differential operators is the fact that the sequence of symbols associated with it is exact if the initial operator p is elliptic (Quillen, Sweeney [137], whereas the analogous statement is not true for the first Spencer resolvent, even in the simplest cases, for example when the operator p is equal to $\bar{\partial}$. On the other hand, the cohomologies of the second resolvent for sufficiently large k are isomorphic to the cohomologies of the first resolvent.

Spencer [129] verified the exactness of the second resolvent in the analytic case, i.e., in the case when all the objects X, E, F, and p are real-analytic. By the foregoing this result is equivalent to local solvability in the class of convergent power series of any regular system with analytic coefficients (see also [83]). Spencer's reasoning was based on a certain lower estimate for the mappings in his δ-sequence, but the proof given by Spencer for the inequality [129] contains a flaw. The flaw was remedied in the work of Ehrenpreis [76], Guillemin and Sterberg [85], and Sweeney [135]. Notable among the papers in which this result has been applied to differential geometry are [130] and [85].

It is essential to point out that the local solvability problem in the analytic case was studied in the early part of the century by Riquier, who also considered nonlinear systems of equations. The conditions guaranteeing solvability in his theory, namely passivity and orthonomy, are useful for verification in specific cases, but they are scarcely applicable to the general theory. In this sense the Spencer regularity condition is more suitable; in particular, it can be written in simple algebraic form (see below). The fact of local solvability of a regular system and, hence, the exactness of the Spencer resolvents in the analytic case can be verified by classical means, specifically the repeated application of the Cauchy-Kovalevskaya theorem.

By a simple device attributable to Grothendieck, Spencer [129] infers from the exactness of the second resolvent a particular result, which may be stated as follows in application to the local solvability problem. If p is a regular differential operator with analytic coefficients, defined in the neighborhood of the coordinate origin of the space R^n, then system (3.1) is solvable in the class of germs over the point $x = 0$ of C^∞-functions under the condition that the differential operator $DD^* + D^* D: \widetilde{C}^1_k \rightarrow \widetilde{D}^1_k$ is elliptic; here D^* is the formal adjoint of the operator D.

Beyond this result little is known concerning the local solvability problem for the class of infinitely-differentiable functions. We do know, however, that there exist scalar operators for which the equation $pu = f$ is unsolvable even locally in the class of generalized functions for the "majority" of smooth functions f (the compatibility condition for scalar systems is empty). The first example of such an operator was found by Hans Lewy, the total class of such operators being isolated later by Hörmander; the simplest operator R^2 has the form $\partial/\partial x_1 + ix_1\partial/\partial x_2$. In all the examples the differential operators satisfy the Spencer regularity condition, because for scalar operators this condition connotes only degeneracy of the principal part.

Matzumura [115] defined a class of nonquadratic systems of first-order equations having an analogous property. Hörmander [88] found what is essentially the broadest class of first-order systems with one unknown function for which local solvability holds. The cited paper of Hörmander contains a more precise global result; we shall discuss it in §4.

Palamodov has studied the local solvability problem in the analytic case in [38]. He correctly formulates a solvable initial-value problem of the Cauchy-Goursat-Riquier type for a system of general form with analytic coefficients. In [38] algebraic constructions are used that generalize the formulations of §2.† Let $D_k(E)$

†See also B. Malgrange, Cohomologie de Spencer (d'apres Quillen), Publ. Sémin. Math. d'Orsay (1966).

be the bundle of differential operators of order no higher than k, acting from E into the trivial one-dimensional bundle I [note that it is equal to Hom $(J^k(E), I)$, i.e., is dual relative to the jet bundle], and let D (E) be the limit of the direct spectrum $\{D_k(E)\}$. The operator p corresponds to the bundle map

$$\widetilde{D(F)} \xrightarrow{p'} \widetilde{D(E)}, \tag{3.7}$$

which acts according to the formula $a \to ap$. As noted by Lopatinskii [29], the fiber of the sheaf $\widetilde{D(I)}$ is a Noetherian left ring. Inasmuch as over every point $x \in X$ map (3.7) is a mapping of left $\widetilde{(DI)}_x$ modules of finite type, it is inferred from the Noetherian character of the ring $\widetilde{D(I)}_x$ that there exists in the neighborhood of x a differential operator $q: \widetilde{F} \to \widetilde{G}$ such that the map $q': \widetilde{D(G)} \to D(F)$ is equal to the kernel of p'. We find, therefore, the operator q, which realizes the compatibility condition (3.2) for any system (3.1) with analytic coefficients.

Next we consider the sheaf M equal to the cokernel of map (3.7); we note that in the event $X = R^n$ and the coefficients of p are constant this sheaf is constant, and its fiber is isomorphic to the module M described in § 2. In the sheaf M we introduce the filtering $\{M_k\}$ equal to the image of the filtering $\{\widetilde{D_k(E)}\}$ in $\widetilde{D(E)}$, then we construct the graded sheaf $\mathrm{gr}\, M = \oplus M_k/M_{k-1}$ of modules over the sheaf of local rings $\widetilde{I}$. The following condition L_x on the operator p is formulated in these terms: The $\widetilde{I}$ -module $\mathrm{gr}\, M_x$ is free. This condition is similar to the regularity condition on p in the neighborhood of x. A certain algebraic procedure applied to the module $\mathrm{gr}\, M_x$ leads to the statement of the desired initial-value problem, which has the form

$$a_i(D)\, u\,|_{L_i} = v_i, \quad i = 1, \ldots, N. \tag{3.8}$$

Here a_i represents certains differentizl operators with constant coefficients, L_i are coordinate subspaces in an arbitrary preselected coordinate system in the neighborhood of x, and v_i are arbitrarily-specified convergent power series of the appropriate groups of variables. The fundamental result of [38] states that under fulfillment of condition L_x system (3.1) has one and only one solution $u \in E_x$ satisfying (3.8) for any right-hand side $f \in F$ such that $qf = 0$. In the case when the operator p is scalar, condition L_x implies that the principal part of that operator is not equal to zero at x; in this case (3.7) gives the initial conditions of the Cauchy problem or the Goursat problem, depending on whichever coefficients of the principal part of p are nonvanishing at x. In the general case the subspaces L_i can have any dimension from 0 to dim X; the set of initial conditions (3.7) is empty if and only if there is a differential operator r: $\tilde{F}_x \to \tilde{E}_x$ such that the composition rp is the identity operator.

The following assertion is also proved in [38]: For any differential operator p: $\tilde{E} \to \tilde{F}$ (all structures are, as before, analytic) and point $x \in X$ there exists a neighborhood U and proper analytic subset $S \subset U$ such that the $\tilde{I}_x$-module gr M_x is free at all points $x \in U \setminus S$, i.e., fulfillment of condition L_x is the rule rather than the exception.

The foregoing results are used in Palamodov's paper [40], which is concerned with differential operators in coherent analytic sheaves. Let K be a field, let A be a K-algebra, and let Φ and Ψ be A-modules. According to I. M. Gel'fand, the K-linear map $\alpha: \Phi \to \Psi$ is said to be an A-differential operator of order no higher than k if for any elements $b_0, \ldots, b_k \in A$ the equation $(\mathrm{ad} b_0) \ldots (\mathrm{ad} b_k)\alpha = 0$ is true. The associated action of an element $b \in A$ on the linear map α is determined here according to the formula $(\mathrm{ad} b)\alpha = \alpha b - b\alpha$. Let Ω be a domain in $\mathbf{C}^n$, let H be the sheaf of germs of holomorphic functions in Ω, and let **F** and **E** be coherent **H**-sheaves. The map α: **F** $\to$ **E** of sheaves of **C**-linear spaces is called an **H**-differential operator if over overy point $x \in \Omega$ the fiber α_x of that map is an H_x-differential operator. If the sheaves **F** and **E** are free and of finite type, this definition coincides with the usual definition of the differential operator with analytic coefficients. Palamodov shows that for any coherent sheaves **F**, **G** $\subset$ **F** and point $x \in \Omega$ there exists in the neighborhood of x a differential operator acting from **F** into the direct sum of structural sheaves of certain analytic subsets of Ω (essential analytic sets of the factor sheaf **F**/**G**), the kernel of which is equal to **G**. There also exists an algebraic condition on the **H**-differential operator α: **F** $\to$ **E** which is necessary and sufficient in order for the kernel of that operator to be a coherent subsheaf of the sheaf **F**. We note that differential operators of this type occur in the theorem on the exponential representation of the solutions of system of equations with constant coefficients (see § 2).

§ 4. The Global Solvability Problem and the Generalized Neumann Problem

The problem area associated with the generalized Neumann problem had its origin in the work of Kodaira, who formulated, analogous to the Hodge decomposition, an orthogonal decomposition connected with the operator $\bar{\partial}$ on a compact complex-analytic manifold. The attempt to extend Kodaira's theory to finite complex-analytic manifolds* leads to a boundary-value problem called the $\bar{\partial}$-Neumann problem, which was formulated in 1955 by Spencer. Essentially it embodies the following. Let Ω be a finite submanifold of a complex-analytic manifold X, and let ω be its boundary. Consider the space $A = \oplus A^{p,q}$ of smooth differential forms on Ω, where A is graded on the "holomorphic" and "antiholomorphic" degrees for which the L_2-norm determined by means of a certain Hermitian metric on X is finite. Also, let δ: A → A be the differential operator formally adjoint to $\bar{\partial}$. We introduce the Laplace operator $\Box = \bar{\partial}\delta + \delta\bar{\partial}$ with domain of definition $\mathbf{D}_{\Box}$ generated by forms $u \in A$ satisfying the following boundary conditions on ω:

$$\begin{aligned} (\bar{\partial}v,\ u) &= (v,\ \delta u), \\ (\bar{\partial}v,\ \bar{\partial}u) &= (v,\ \delta\bar{\partial}u), \end{aligned} \qquad \forall v \in A. \tag{4.1}$$

The $\bar{\partial}$-Neumann problem is essentially to establish the existence of an operator N : A → A, bounded under the L_2-norm, of zero degree relative to grading in A such that for any form $u \in A$ the form Nu belongs to $D_{\Box}$ and the following orthogonal decomposition holds:

$$u = \Box Nu + Hu,$$

in which H is the projector onto the space $\operatorname{Ker}\Box$ of harmonic forms, where $HN = NH = \bar{\partial}N - N\bar{\partial} = 0$.

The only case in which the boundary condition (4.1) in this problem is not a condition of the Shapiro–Lopatinskii type is when the $\bar{\partial}$-Neumann problem is treated for the forms of maximal "antiholomorphic" degree (see [121]). Consequently, despite the fact that the operator $\Box$ itself is elliptic, the solvability of the $\bar{\partial}$-Neumann

*We recall that such a manifold is interpreted as an open relatively-compact subset of a complex-analytic manifold whose boundary is a smooth real manifold of codimension one.

problem depends strongly on the manifold Ω, in particular on the geometry of its boundary.

One version of the $\bar{\partial}$-Neumann problem was investigated earlier by Garabedian and Spencer [80]. Kohn and Spencer [99] investigated the analogous problem for an operator comprising a linear combination of ∂ and $\bar{\partial}$ and solved it in special cases. Morrey [120] solved the $\bar{\partial}$-Neumann problem for forms of the type (0, 0) and (0, 1) on manifolds of a special type and used the result to prove that every compact real-analytic manifold can be analytically embedded in a Euclidean space. His method is based on the inequality

$$c\int_{\omega} |u|^2 ds \leqslant Q(u, u) + \|u\|^2 \tag{4.2}$$

for differential forms u of positive "antiholomorphic" degree satisfying the first of the boundary conditions (4.1). Here $|u|$ is the point norm of u, and Q (u, u) is the Dirichlet integral for the operator ∂, i.e., the sum $\|\bar{\partial}u\|^2 + \|\delta u\|^2$. Morrey's method relies heavily on the following regularity theorem: If the form u is regular out to the boundary of Ω, the form Nu has the same property; however, the proof of this theorem is not complete in Morrey's paper. This flaw was remedied by Kohn in [92, 93], in which he simultaneously verified the solvability of the $\bar{\partial}$-Neumann problem on any finite manifold with a pseudoconvex boundary and proved for such manifolds the finite-dimensionality of the harmonic space $H^{p,q}$ for $q > 0$. From this result Kohn deduced anew the solution obtained earlier by Grauert for the Lewy problem. In the same paper Kohn obtained a new proof of the Newlander-Nirenberg theorem on the existence of holomorphic coordinates on an integrable almost-complex manifold. Morrey [121] somewhat simplified Kohn's proof of the regularity theorem. Ash [59] proposed a simpler derivation of the Morrey-Kohn inequality (4.2) on the basis of the moving frame method.

In [94] Kohn translates certain results of harmonic theory to odd-dimensional real compact manifolds. He introduces the notion of the integrable almost-complex structure on such a manifold and constructs an orthogonal decomposition for the operator $\bar{\partial}$ corresponding to that structure, assuming that the manifold satisfies a certain condition of the pseudoconvexity type. The proof of the decomposition is based on the estimate

$$c\|u\|_{1/2}^2 \leqslant Q(u, u) + \|u\|^2, \tag{4.3}$$

in which Q is the Dirichlet integral for $\bar{\partial}$. In this inequality the constant 1/2 cannot be increased. Later Hörmander [88], working from the above paper of Kohn, investigated the class of what he called subelliptic differential operators, for which there is an a priori estimate with a 1/2 "loss of smoothness," as opposed to the a priori estimate for elliptic operators. Thus, the first example of a subelliptic operator was the Laplacian on the odd-dimensional manifolds investigated by Kohn.

In [98] Kohn and Rossi use the method of harmonic integrals to prove the following theorem: If Ω is a finite complex-analytic manifold whose boundary is connected and the Lewy form of the boundary has at least one eigenvalue at every point, then function on the boundary that satisfies the "tangential" Cauchy-Riemann equations can be extended onto Ω as a holomorphic function. Also in [98] Kohn and Rossi find sufficient conditions on the boundary of Ω in order for the analogous extension theorem to be valid for $\bar{\partial}$-closed differential forms.

Androtti and Vesentini [57] have investigated the cohomologies of a complex-analytic manifold X without boundary, using the following Carleman-type inequality for this purpose:

$$c\|u\|_{\varphi} \leqslant \|\bar{\partial} u\|_{\varphi} + \|\delta_{\varphi} u\|_{\varphi}. \tag{4.4}$$

Here $\|\cdot\|_{\varphi}$ is the norm* corresponding to the scalar product

$$(u, v)_{\varphi} = \int_X \exp(\varphi)\, u\bar{v}\, dx,$$

δ_{φ} is the adjoint of the operator $\bar{\partial}$ with respect to the indicated scalar product, and φ is a function characterizing the structure of X. The manifold X is said to be a-pseudoconvex if there exists on it a real continuous function φ, q-pseudoconvex† outside some compactum K, such that the set of lesser values $\{x \in X, \varphi(x) \leq c\}$ is compact for any c. In [57] the authors deduce a number of theorems of the following kind: If X is q-pseudoconvex and E is a

*Hörmander has used norms of this type (after Carleman) to study the uniqueness of the solution of the Cauchy problem.

†Recall that a function φ specified on a complex manifold is said to be q-pseudoconvex if it is real, belongs to class C^2, and has at every point of its Lewy matrix at least $n-q+1$ positive eigenvalues.

locally-free analytic sheaf on X, then the cohomologies of X with compact supports of differential forms with coefficients from E are finite-dimensional, the number of dimensions not exceeding the number $\dim_C X - q$. This and other allied results of [57] are special cases of the theorems of Andreotti and Grauert [56] concerning arbitrary coherent sheaves on q-pseudoconvex complex spaces.

Hörmander [87] strengthens the Morrey-Kohn inequality (4.2) by application therein of the following Carleman-type norms:

$$\| u \|_{-\varphi} = \int_X \exp(-\varphi) \, | u |^2 dx,$$

in which φ is a plurisubharmonic or, more generally, a pseudoconvex function. The left-hand side of this inequality contains an extra term equal to the norm of u with a certain positive density depending on φ. Hörmander finds a necessary and sufficient condition on φ in order for the given type of inequality to hold for all forms of fixed bidegree (p, q). The strengthened inequality, which in a certain sense represents the dual of inequality (4.4), enables one to deduce a precise theorem for the solvability of the Cauchy-Riemann system of equations in C^n in classes of functions for which the norm $\| \cdot \|_{-\varphi}$ is finite, where φ is a plurisubharmonic function. Theorems of this type are intriguing insofar as they can be used to prove general theorems on systems of equations with constant coefficients; see Malgrange [110], Palamodov [37], and Hörmander [55].

On the basis of the strengthened Morrey-Kohn inequality Hörmander in [87] obtains refined theorems on the density of integral functions with finite norms $\| \cdot \|_{-\varphi}$ in the space of all integral functions in $\mathbf{C}^n$ and proves a special case of the Andreotti-Grauert theorem cited above. An important methodological simplification is injected by Hörmander in his reliance, not on the difficult regularity theorem used by Morrey and Kohn, but on the rather elementary fact that the strong and weak expansions of differential operators coincide.

In [86] Hörmander adapts his method to the investigation of the solvability of a system of first-order equations with the unknown function

$$P_j u = f_j, \; j = 1, \ldots, N, \tag{4.5}$$

in a domain Ω of Euclidean space. Assuming that the commutator of any two operators P_j and P_k are linearly expressed in terms of

$P_1, \ldots, P_N$, Hörmander writes the compatibility conditions in the form of relations $P_j f_k = P_k f_j$ up to zero-order terms on $f_1, \ldots, f_N$. Another condition bears on the principal parts p_j of the operators P_j. It states that any commutator of the form $[p_j, \bar{p}_k]$ (the coefficients of $\bar{p}_k$ are the complex conjugates of the coefficients of p_k) is a linear combination of the operators p_i and $\bar{p}_i$. This type of condition is necessary, as otherwise any equation of system (4.5) could be independently unsolvable, even locally (see §3). Under these and certain additional rather minor restrictions Hörmander proves the solvability of system (4.5) in class $L_2^{loc}(\Omega)$ for any domain Ω pseudoconvex with respect to that system. The latter condition implies that there exists in Ω a real function φ, satisfying at every point a certain condition interconnecting its second derivatives, i.e., a condition of the Lewy type, such that any set of lesser values of φ is compact. This theorem affords a sufficiently complete solution of the local solvability problem for first-order systems of the form (4.5); it includes Nirenberg's result of [124] as a special case.

Conceptually speaking, Kohn and Nirenberg's paper [97] is a continuation of Kohn's work [92-94]. In it they investigate an arbitrary first-order system of equations $Lu = f$ in a domain $\Omega \subset \mathbf{R}^n$. In addition to the operator L they consider a certain first-order differential operator M such that $ML = 0$. The problem analyzed by the authors may be stated in a form analogous to the $\bar{\partial}$-Neumann problem: For every vector function u smooth in Ω construct the orthogonal decomposition

$$u = LL^*Nu + M^*MNu + Hu, \tag{4.6}$$

in which H is the projector onto the subspaces

$$\mathbf{H} = \{u \in D_{L^*},\ L^*u = Mu = 0\},$$

and N is an operator, bounded under L_2-norms, which preserves smoothness such that $HN = NH = 0$ and that $Mu = 0$ implies $MNu = 0$. This existence of such a decomposition implies the solvability of the system $Lu = f$ under the conditions $Mf = 0$ and $f \perp \mathbf{H}$. Kohn and Nirenberg reduce the existence problem for such a decomposition to solvability of the equation

$$Q(u, v) = (\alpha, v), \tag{4.7}$$

where α is a given vector function, $Q(u, v) = (L^*u, L^*v) + (Mu, Mv)$, and v runs through the set of smooth elements of the domain of definition of the operator L^*. The investigation is concerned primarily with the solvability of Eq. (4.7) for a certain broader class of bilinear forms Q. The authors give several versions of conditions guaranteeing the solvability of Eq. (4.7) and the regularity of its solution out to the boundary of Ω. The chief condition is contained in the fact that the norm $|Q(u, u)|^{1/2}$ must be compact under the L_2-norm. For the $\bar{\partial}$-Neumann problem on a pseudoconvex manifold this condition is a consequence of the Morrey-Kohn inequality (4.2), while for the analogous problem on an odd-dimensional manifold it follows from inequality (4.3).

In [88] Hörmander, as part of an investigation of subelliptic operators, finds an extensive class of quadratic forms Q for which the inequality $\|u\|^2_{1/2} \leq cQ(u, u)$ holds. In application to the problem of the decomposition (4.6), however, his results produce rather obscure conditions on the operators L and M. Some later results bearing on the problem (4.6) are discussed in Kohn's paper [96].

A general technique for handling the global solvability problen for systems of equations of arbitrary order has been proposed by Spencer [129-133]. Proceeding from (3.6), Spencer formulates the following generalized Neumann problem. Let Ω be a finite submanifold of a differentiable manifold X on which is given a regular operator p. Also, let A be the space of smooth cross sections of the bundle $C_k = \oplus C_k^i$ over Ω, and let $D^* : A \to A$ be a differential operator formally adjoint to D with respect to the scalar product on A defined by some Hermitian metric on C_k. Consider the operator $\Box = DD^* + D^*D$ with domain of definition $D_\Box$ formed by cross sections $u \in A$ satisfying the following boundary conditions on ω:

$$\begin{aligned} (Dv, u) &= (u, D^*u), \\ (Dv, Du) &= (v, D^*Du), \end{aligned} \quad \forall v \in A. \tag{4.8}$$

The D-Neumann problem calls for the construction of an operator $N: A \to A$, bounded under L_2-norms and of zero degree under grading in A, such that for any $u \in A$ it is true that $Nu \in D_\Box$ and $u = \Box Nu + Hu$, where H is the projector onto $\operatorname{Ker}\Box$; here $HN = NH = DN - ND = 0$. It is readily perceived that if this problem is solvable the "harmonic" space $\operatorname{Ker}\Box$ is isomorphic to the D-cohomology of A.

Spencer announced in [133] that for any regular elliptic operator p the D-Neumann problem is solvable and the harmonic space is trivial in positive dimensions in any sufficiently small convex domain. Spencer formulated analogous assertions in [131, 132]. The formal reduction to which we allude in §3 shows that the foregoing assertion implies the local solvability of any regular elliptic system (3.1). However, the proof of Spencer's assertion has not been published and, besides, the local solvability problem for regular elliptic systems is involved in diverse forms in a number of unsolved problems posed at the 1968 summer seminar of the American Mathematical Society. Consequently, the above assertions of Spencer must be treated as hypotheses.

A paper by Sweeney [137] is devoted to the investigation of the D-Neumann problem. Using the above results of Kohn and Nirenberg, he reduces the problem to proof of the inequality

$$c\|u\|_r^2 \leqslant Q(u,u) + \|Bu\|_\omega^2, \quad r>0,$$

in which Q (u, u) is the Dirichlet integral for the operator D and B is a boundary differential operator, the kernel of which coincides with the space of cross-sections satisfying the first condition (4.8). Using Hörmander's method of [88], Sweeney reduces the above inequality to the analysis of certain pseudodifferential operators on ω, arriving ultimately at an algebraic condition sufficient for the solvability of the D-Neumann problem. This condition, however, scarcely qualifies as necessary. As the author points out, it does not hold for the operator $\bar\partial$; see also [136]. In the same paper Sweeney constructs a certain complex of sheaves on ω, which might aptly be termed the "tangential" second resolvent of Spencer. For $\bar\partial$ this construction leads to the "tangential" operator $\bar\partial$ studied by Kohn and Rossi in [98] and by Kohn in [94]. Using the subellipticity criteria deduced by Hörmander [88], the author obtains a new proof of Kohn's theorem [94] on the solvability of the $\bar\partial$-Neumann problem on an odd-dimensional manifold.

In [66] Cenkl seeks sufficient conditions on the regular elliptic operator p: $\widetilde{E} \to \widetilde{F}$ in order for the cohomologies of the sheaf $\theta = \operatorname{Ker} p$ to be zero in positive dimensions. Assuming that the second resolvent of Spencer for p is exact, Cenkl reduces the problem to an investigation of the D-Neumann problem for the so-called β-resolvent, which is a modification of the second resolvent of Spencer. Transforming the Dirichlet integral associated with the β-resolvent, Cenkl isolates a leading term in the guise of a certain bilinear form, which is expressed fairly simply in terms of the coefficient of the operator and the geometry of the bundle E. The condition of "sufficient" positivity of this form guarantees the triviality of the cohomologies $H^i(\Omega, \theta)$, $i > 0$, for the case of a compact manifold Ω. In the special case $p = \bar{\partial}$ the latter assertion reverts to the well-known theorem of Kodaira [90]. Cenkl deduces an analogous "vanishing" theorem for a finite complex-differentiable manifold Ω, imposing on its boundary a condition similar to the pseudoconvexity condition.

Literature Cited

1. Agranovich, M. S., Elliptic singular integrodifferential operators, Usp. Mat. Nauk, 20(5): 3-120 (1965).
2. Agranovich, M. S., Positive boundary-value problems for certain first-order systems, Trudy Moskov. Mat. Obshch., 16: 3-24 (1967).
3. Agranovich, M. S., On the theory of boundary-value problems for symmetrizable first-order systems, Mat. Sb., 73(2): 161: 197 (1967).
4. Agranovich, M. S. and Dynin, A. S., General boundary-value problems for elliptic systems in a multidimensional domain, Dokl. Akad. Nauk SSSR, 146(3): 511-514 (1962).
5. Vainberg, B. R. and Grushin, V. V., Uniformly-nonelliptic problems (I), Mat. Sb., 72(4): 602-636 (1967).
6. Vainberg, B. R. and Grushin, V. V., Uniformly-nonelliptic problems (II), Mat. Sb., 73(1): 126-154 (1967).
7. Vishik, M. I. and Éskin, G. I., Elliptic equations in convolutions in a bounded domain and their applications, Usp. Mat. Nauk, 22(1): 15-16 (1967).
8. Volevich, L. R., On general systems of differential equations, Dokl. Akad. Nauk SSSR, 132(1): 20-23 (1960).
9. Volevich, L. R., Regularity of the solutions of systems of differential equations with variable coefficients, Usp. Mat. Nauk, 16(2): 163-164 (1961).
10. Volevich, L. R., Solvability of boundary-value problems for general elliptic systems, Mat. Sb., 68(3): 373-416 (1965).
11. Volevich, L. R., Local properties of inhomogeneous pseudodifferential operators, Trudy Moskov. Mat. Obsch., 16: 51-98 (1967).
12. Vostretsov, B. A., Structure of the analytic solutions of a class of systems of linear partial differential equations with constant coefficients, Dokl. Akad. Nauk SSSR, 161(6): 1259-1262 (1965).
13. Galakhov, M. A., Nonclassical boundary-value problems for symmetric linear first-order partial differential systems, Differentsial'nye Uravneniya, 1(12): 1620-1627 (1965).
14. Gel'fand, I. M. and Shilov, G. E., Generalized Functions, Part 3: Aspects of the Theory of Differential Equations, Fizmatgiz, Moscow (1958).
15. Gluzberg, E. I., On the Cauchy problem for a countable system of partial differential equations, Proc. First Kazakhstan Sci. Conf. Mathematics and Mechanics, Alma-Ata, 1963, Nauka (1965), pp. 55-57.

16. Golets, B. I. and Verenich, I. I., The Cauchy problem for partial differential systems with variable coefficients, Abstracts of Papers of the Twelfth Scientific Meeting of the Physical and Mathematical Sciences Section of Chernovitsy University, Chernovitsy (1966), pp. 127-129.
17. Golets, B. I. and Eidel'man, S. D., Nine theorems on I. G. Petrovskii-correct systems, Dopovidi Akad. Nauk Ukrain. RSR, No. 9, pp. 1106-1111 (1966).
18. Gorin, E. A., Solvability of the Cauchy problem, Vest. Moskov. Univ., Mat. Mekh., No. 4, pp. 6-12 (1965).
19. Dezin, A. A., Existence and uniqueness theorems for the solutions of boundary-value problems for partial differential equations in functional spaces, Usp. Mat. Nauk, 16(3): 21-73 (1959).
20. Didenko, V. P., Solvability of certain boundary-value problems for elliptic systems with degenerate order at the boundary, Differentsial'nye Uravneniya, 3(1): 11-18 (1967).
21. Zhitarashu, N. V., A priori estimates and the solvability of general boundary-value problems for general elliptic systems with discontinuous coefficients, Dokl. Akad. Nauk SSSR, 165(1): 24-27 (1965).
22. Zhitarashu, N. V., Schauder estimates and the solvability of general boundary-value problems for general parabolic systems with discontinuous coefficients, Dokl. Akad. Nauk SSSR, 169(3): 511-514 (1966).
23. Zolotareva, E. V., The Dirichlet problem for some second-order elliptic systems, Differentsial'nye Uravneniya, 3(1): 59-68 (1967).
24. Ivasishen, S. D. and Lavrenchuk, V. P., Solvability of the Cauchy problem and certain boundary-value problems for general parabolic systems in a class of increasing functions, Depovidi Akad. Nauk Ukrain. RSR, A, No. 4, pp. 299-303 (1967).
25. Kovach, Yu. I., Application of the theorem of differential inequalities to the Goursat problem for a linear system of partial differential equations, Differentsial'nye Uravneniya, 1(3): 411-420 (1965).
26. Kopaček, J. and Suha, M., The Cauchy problem for weakly-hyperbolic systems of linear differential equations with constant coefficients, Časop. Peštov. Mat., 91(4): 431-452 (1966).
27. Kuz'min, E. M., The Dirichlet problem for elliptic systems in a space, Differentsial'nye Uravneniya, 3(1): 155-157 (1967).
28. Lëdnev, N. A., A new method of solving partial differential equations, Mat. Sb., 22: 205-266 (1948).
29. Lopatinskii, Ya. B., Properties of linear differential operators, Mat. Sb., 17(2): 267-285 (1945).
30. Lopatinskii, Ya. B., Reduction of a system of differential equations to canonical form, in: Theory and Applications of Mathematics, No. 2, Lvov. Univ., Lvov (1963), pp. 53-64.
31. Matiichuk, M. I. and Éidel'man, S. D., Parabolic systems with coefficients satisfying the Dini condition, Dokl. Akad. Nauk SSSR, 165(3): 482-485 (1965).
32. Matiichuk, M. I. and Éidel'man, S. D., On the fundamental solutions of elliptic systems, Ukrain. Mat. Zh., 18(2): 22-41 (1966).
33. Nguen Tkhya Khop, The Dirichlet problem for strongly-connected elliptic systems, Dokl. Akad. Nauk SSSR, 171(2): 292-295 (1966).

34. Oskolkov, A. M., A priori estimates of the first derivatives for two-dimensional linear strongly-elliptic systems and elliptic mappings, Trudy Mat. Inst. Akad. Nauk SSSR, 92:182-191 (1966).

35. Palamodov, V. P., Construction of polynomial ideals and their factor spaces in spaces of infinitely-differentiable functions, Dokl. Akad. Nauk SSSR, 141(6): 1302-1305 (1961).

36. Palamodov, V. P., Systems of differential equations with constant coefficients, Dokl. Akad. Nauk SSSR, 148(3):523-526 (1963).

37. Palamodov, V. P., Linear Differential Operators with Constant Coefficients, Nauka, Moscow (1967).

38. Palamodov, V. P., Differential operators in the class of convergent power series and the Weierstrass preparation theorem, Funktsional. Analiz i Ego Prilozhen., 2(3):58-69 (1968).

39. Palamodov, V. P., Remark on the exponential representation of the solutions of differential equations with constant coefficients, Mat. Sb., 75(3):417-434 (1968).

40. Palamodov, V. P., Differential operators in coherent analytic sheaves, Mat. Sb., 70(3):390-422 (1966).

41. Paneyakh, B. P., On general systems of differential equations with constant coefficients, Dokl. Akad. Nauk SSSR, 138(2):297-300 (1961).

42. Parasyuk, L. S., Generalized fundamental solution of elliptic systems of differential equations with discontinuous coefficients, Ukrain. Mat. Zh., 18(4): 124-126 (1966).

43. Roitberg, Ya. A. and Sheftel', Z. G., Boundary-value problems with parameter in L_p for Douglis-Nirenberg-elliptic systems, Ukeain. Mat. Zh., 19(1):115-120 (1967).

44. Skripnik, I. V., The α-Neumann problem, Dopovidi Akad. Nauk Ukrain. RSR, No. 3, pp. 295-299 (1966).

45. Solomyak, M. Z., On first-order linear elliptic systems, Dokl. Akad. Nauk SSSR, 150(1):48-51 (1963).

46. Solonnikov, V. A., General boundary-value problems for systems elliptic in the Douglis-Nirenberg sense (I), Izv. Akad. Nauk SSSR, Ser. Mat., 28(3):665-706 (1964).

47. Solonnikov, V. A., General boundary-value problems for systems elliptic in the Douglis-Nirenberg sense (II), Trudy Mat. Inst. Akad. Nauk SSSR, 92:233-297 (1966).

48. Sochneva, V. A., On the Cauchy problem for systems of partial differential equations with variable coefficients, Differentsial'nye Uravneniya, 2(11):1520-1530 (1966).

49. Sochneva, V. A., Solutions of general linear systems of partial differential equations analytic on one variable, Izv. Vuzov, Mat., No. 2, pp. 67-73 (1967).

50. Tovmasyan, N. E., Boundary-value problems for second-order elliptic systems of equations that do not satisfy the condition of Ya. B. Lopatinskii, Dokl. Akad. Nauk SSSR, 160(5):1028-1031 (1965).

51. Tovmasyan, N. E., General boundary-value problem for second-order elliptic systems with constant coefficients (I), Differentsial'nye Uravneniya, 2(1):3-23 (1966).

52. Tovmasyan, N. E., General boundary-value problem for second-order elliptic systems with constant coefficients (II), Differentsial'nye Uravneniya, 2(2): 163-171 (1966).
53. Fridlender, V. R., Polynomial matrices and systems of partial differential equations, Izv. Vuzov, Mat., No. 5, pp. 118-123 (1966).
54. Fridlender, V. R. and Khamitov, L. Kh., The generalized Duff-Friedman problem, Izv. Vuzov, Mat., No. 5, pp. 101-107 (1967).
55. Hörmander, L., An Introduction to Complex Analysis in Several Variables, Van Nostrand, Princeton, N. J. (1966).
56. Andreotti, A. and Grauert, H., Finiteness theorems for the cohomologies of complex spaces [in French], Bull. Soc. Math. France, 90:193-259 (1962).
57. Andreotti, A. and Vesentini, E., Carleman estimates for the Laplace-Beltrami equations on complex manifolds, Publs. Math. Inst. Heutes Études Sci., No. 25, pp. 313-362 (1965).
58. Apostolatos, N. and Kulisch, U., Integrability and integration of overdetermined systems of partial differential equations [in German], Z. Angew. Math. und Mech., 47(4):261-268 (1967).
59. Ash, M. E., The basic estimates of the $\bar{\delta}$-Neumann problem in the non-Kählerian case, Amer. J. Math., 86(2):247-254 (1964).
60. Atiyah, M. F. and Bott, R., The index problem for manifolds with boundary, in: Differential Analysis, Oxford Univ. Press, London (1964), pp. 175-189.
61. Avantaggiati, A., On the fundamental principal matrices for a class of elliptic and hypoelliptic systems of differential equations [in Italian], Ann. Mat. Pura Appl., 65:191-237 (1964).
62. Avantaggiati, A., New contributions to the investigation of the convolution problem for first-order elliptic systems [in Italian], Ann. Mat. Pura Appl., 69:107-169 (1965).
63. Brenner, P., The Cauchy problem for symmetric hyperbolic systems in L_p, Math. Scandinav., 19(1):27-37 (1966).
64. Caldwell, W. V., Some relationships between Bers and Beltrami systems and linear elliptic systems of partial differential equations, Canad. J. Math., 17(4): 627-642 (1965).
65. Cenkl, B., The elliptic differential operators, Comment. Math. Univ. Carolinae, 8(2):175-197 (1967).
66. Cenkl, B., Vanishing theorem for an elliptic differential operator, J. Differential Geom., 1:381-418 (1967).
67. Chen Liang-jin, The Dirichlet problem for a class of systems of degenerate elliptic equations, Chinese Math., 5(3):409-417 (1964).
68. Chou Nien-tei, A function theory for a system of first-order elliptic partial differential equations, Sci. Abstr. China, Math. and Phys. Sci., 2(4):9-10 (1964).
69. Cinquini, C. M., Systems of partial differential equations in several independent variables [in Italian], Semin. 1962-1963 Analisi, Algebra, Geometria e Topol., Vol. 1, Rome (1965), pp. 101-122.
70. Dombrowski, P., Maximal explicit solutions (Riemann surfaces) of the Cauchy initial-value problem for first-order partial differential equations for one unknown function on C^∞-manifolds (I) [in German], Math. Ann., 160(3):195-232 (1965).

71. Dombrowski, P., Maximal explicit solutions (Riemann surfaces) of the Cauchy initial-value problem for first-order partial differential equations for one unknown function on C^{∞}-manifolds (II) [in German], Math. Ann., 160(4): 257-279 (1965).
72. Dombrowski, P., Maximal explicit solutions (Riemann surfaces) of the Cauchy initial-value problem for first-order partial differential equations for one unknown function on C^{∞}-manifolds (III) [in German], Math. Ann., 161(1): 26-66 (1965).
73. Douglis, A. and Nirenberg, L., Interior estimates for elliptic systems of partial equations, Commun. Pure Appl. Math., 8(4): 503-538 (1955).
74. Douglis, A., Some applications of the theory of distributions to several complex variables, Seminar on Analytic Functions, Vol. I, Inst. Adv. Study, Princeton (1958), pp. 65-79.
75. Ehrenpreis, A fundamental principle for systems of linear differential equations with constant coefficients and some of its applications, Proc. Internat. Symposium Linear Spaces, Jerusalem, 1960, Acad. Press, Jerusalem; Pergamon, New York (1960).
76. Ehrenpreis, L., Guillemin, V. W. and Sternberg, S., On Spencer's estimate for δ-Poincaré, Ann. Math., 83(1): 128-138 (1965).
77. Foias, C. and Gussi, G., A uniqueness theorem for the solution of the Cauchy problem for certain linear systems of partial differential equations [in French], Atti Acad. Naz. Lincei. Rend. Cl. Sci. Fis. Mat. Natur., 29(6): 509-514 (1960).
78. Friedrichs, K. O. and Lax, P. D., Boundary-value problems for first-order operators, Commun. Pure Appl. Math., 18: 355-388 (1965).
79. Fuglede, B., A priori inequalities connected with systems of partial differential equations, Acta Math., 105 (1961).
80. Garabedian, P. R. and Spencer, D. C., Complex boundary problems, Trans. Amer. Math. Soc., 73: 223-242 (1952).
81. Garding, L., Energy inequalities for hyperbolic systems, in: Differential Analysis, Oxford Univ. Press, London (1964), pp. 209-225.
82. Garding, L., A variation on the Cauchy majorizing method [in French], Acta Math., 114(1-2): 143-158 (1965).
83. Goldschmidt, H., Existence theorems for analytic linear partial differential equations, Ann. Math., 86(2): 246-270 (1967).
84. Guillemin, V. W., Some algebraic results concerning the characteristics of overdetermined partial differential equations, Amer. J. Math., 90(1): 270-284 (1968).
85. Guillemin, V. W. and Sternberg, S., An algebraic model of transitive differential geometry, Bull. Amer. Math. Soc., 70(1): 16-47 (1964).
86. Hörmander, L., The Frobenius-Nirenberg theorem, Arkiv Mat., 5(5): 425-432 (1964).
87. Hörmander, L., L^2-estimates and existence theorems for the δ operator, Acta Math., 113(1-2): 89-152 (1965).
88. Hörmander, L., Pseudodifferential operators and nonelliptic boundary problems, Ann. Math., 83(1): 129-209 (1966).
89. Hua Loo-Keng and Wu Tze-chien, On the uniqueness theorem of the Dirichlet problem of the linear elliptic system, Sci. Abstr. China. Math. and Phys. Sci., 3(3): 9-10 (1965).

90. Kodaira, K., On a differential geometric method in the theory of analytic stacks, Proc. Nat. Acad. Sci. USA, 39:1268-1273 (1953).

91. Kohn, J. J., Solution of the $\bar{\partial}$-Neumann problem on strongly-pseudoconvex manifolds, Proc. Nat. Acad. Sci. USA, 47:1198-1202 (1961).

92. Kohn, J. J., Harmonic integrals on strongly-pseudoconvex manifolds (I), Ann. Math., 78(1):112-148 (1963).

93. Kohn, J. J., Harmonic integrals on strongly-pseudoconvex manifolds (II), Ann. Math., 79:450-472 (1964).

94. Kohn, J. J., Boundaries of complex manifolds, Proc. Conf. Complex Analysis, Minneapolis, 1964, Springer-Verlag, Berlin (1965), pp. 81-94.

95. Kohn, J. J., Differential operators on manifolds with boundary, in: Differential Analysis, Oxford Univ. Press, London (1964), pp. 57-63.

96. Kohn, J. J., Differential complex, Proc. Internat. Congr. Mathematicians, 1966, Mir, Moscow (1968), pp. 402-409.

97. Kohn, J. J. and Nirenberg, L., Noncoercive boundary-value problems, Commun. Pure Appl. Math., 18:443-492 (1965).

98. Kohn, J. J., and Rossi, H., On the extension of holomorphic functions from the boundary of a complex manifold, Ann. Math., 81(3):451-472 (1965).

99. Kohn, J. J., and Spencer, D. C., Complex Neumann problems, Ann. Math., 66: 98-140 (1957).

100. Komatsu, H., Resolutions by hyperfunctions of sheaves of solutions of differential equations with constant coefficients, Math. Ann., 176(1):77-86 (1968).

101. Kopaček, J., The Cauchy problem for linear hyperbolic systems in L_p, Comment. Math. Univ. Carolinae, 8(3):459-462 (1968).

102. Kuranishi, M., Sheaves defined by differential equations and application to deformation theory of pseudogroup structures, Amer. J. Math., 86:379-391 (1964).

103. Kuranishi, M., Lectures on Involutive Systems of Partial Differential Equations (Publ. Soc. Mat. São Paulo), São Paulo (1967), 77 pages.

104. Lamberti, L., On the Wirtinger system of partial differential equations [in Italian], Rend. Mat. Appl., 23(3-4):419-437 (1964).

105. Lax, P. D. and Phillips, R. S., Local boundary conditions for dissipative symmetric linear differential operators, Commun. Pure Appl. Math., 13:427-455 (1960).

106. Malgrange, B., Division of distributions, I: Extendable distributions [in French], Sém. Schwartz, Fac. Sci. Paris, 1959-60, 4 année, Paris (1960), 22 [1-21] 5.

107. Malgrange, B., Division of distributions, II: The Lojasiewicz inequality [in French], Sém. Schwartz, Fac. Sci. Paris, 1959-60, 4 année, Paris (1960), pp. 22 [1-22] 8.

108. Malgrange, B., Division of distributions, III: The principal theorem [in French], Sém. Schwartz, Fac. Sci. Paris, 1959-60, 4 année, Paris (1960), pp. 23 [1-23] 11.

109. Malgrange, B., Division of distributions, IV: Applications [in French], Sém. Schwartz, Fac. Sci. Paris, 1959-60, 4 année, Paris (1960), pp. 25 [1-25] 5.

110. Malgrange, B., On differential systems with constant coefficients [in French], Colloq. Internat. Centre Nat. Rech. Sci., No. 117, pp. 113-122 (1963).

111. Malgrange, B., Differential systems with constant coefficients [in French], Sém. Bourbaki, Secrét. Math., 1962-63, 15(1): 246 [01-246] 11 (1964).

112. Malgrange, B., Some convexity problems for differential operators with constant coefficients, Sém. Leray, Collège, de France, Paris (1962/63), pp. 190-223.
113. Malgrange, B., Some remarks on the notion of convexity for differential operators, in: Differential Analysis, Oxford Univ. Press, London (1964), pp. 163-174.
114. Mansfield, L. E., A Generalization of the Cartan-Kähler Theorem, Doct. Dissert., Univ. Washington, 1965, 69 pages; Dissert. Abstr., 26(8):4693-4694 (1966).
115. Matzumura, M., Local existence of solutions for certain systems of partial differential equations [in French], Japan J. Math., 32:13-49 (1962).
116. Matsuura, S., On general systems of partial differential operators with constant coefficients, J. Math. Soc. Japan, 13(1):94-103 (1961).
117. Matsuura, S., A remark on ellipticity of general systems of differential operators with constant coefficients, J. Math. Kyoto Univ., 1 (1):71-74 (1961).
118. Milicer-Gruzewska, H., The properties of generalized potentials and a limit problem for elliptic systems [in French], Bull. Acad. Polon. Sci., Sér. Sci. Math. Astron. Phys., 13(2):125-133 (1965).
119. Milicer-Gruzewska, H., On the fundamental solution of eliiptic systems with Hölder coefficients [in French], Bull. Acad. Polon. Sci., Sér. Sci. Mat. Astron. Phys., 13(2):117-124 (1965).
120. Morrey, C. B., The analytic embedding of abstract real-analytic manifolds, Ann. Math., 68:159-201 (1958).
121. Morrey, C. B., The $\bar{\delta}$-Neumann problem on strongly-pseudoconvex manifolds, in: Differential Analysis, Oxford Univ. Press, London (1964), pp. 81-133.
122. Morrey, C. B. and Nirenberg, L., On the analyticity of the solutions of linear elliptic systems of partial differential equations, Commun. Pure Appl. Math., 10:271-290 (1957).
123. Newlander, A. and Nirenberg, L., Complex-analytic coordinates in almost-complex manifolds, Ann., Math., 65:391-404 (1957).
124. Nirenberg, L., A complex Frobenius theorem, Sem. Analytic Functions, Vol. 1, Inst. Adv. Study, Princeton (1957), pp. 172-189.
125. Persson, J., A boundary problem for analytic linear systems with data on intersecting hyperplanes, Math. Scandinav., 14(1):106-110 (1964).
126. Phillips, R. S. and Sarason, L., Singular symmetric positive first-order differential operators, J. Math., Mech., 15(2):235-271 (1956).
127. Schaefer, P. W., On the Cauchy problem for an elliptic system, Arch. Rat. Mech. Anal., 20(5):391-412 (1965).
128. Schechter, M., Systems of partial differential equations in a half-space, Commun. Pure Appl. Math., 17(4):423-434 (1964).
129. Schechter, M., Deformation of structures on manifolds defined by transitive continuous pseudogroups (I), Ann. Math., 76:306-398 (1962).
130. Schechter, M., Deformation of structures on manifolds defined by transitive continuous pseudogroups (II), Ann. Math., 76:399-445 (1962).
131. Schechter, M., Harmonic integrals and Neumann problems associated with linear partial differential equations, Outlines of the Joint Soviet-American Symposium on Partial Differential Equations, Novosibirsk (1963), pp. 253-260.

132. Spencer, D. C., Existence of local coordinates for structures defined by elliptic pseudogroups, in: Differential Analysis, Oxford Univ. Press, London (1964), pp. 135-162.

133. Spencer, D. C., De Rham theorems and Neumann decompositions associated with linear partial differential equations, Ann. Inst. Fourier, 14(1):1-19 (1964).

134. Spencer, D. C., Deformation of structures on manifolds defined by transitive continuous pseudogroups (II), Ann. Math., 81:389-450 (1962).

135. Sweeney, W. J., The δ-Poincaré estimate, Pacific J. Math., 20(3):559-570 (1967).

136. Sweeney, W. J., A noncompact Dirichlet norm, Proc. Nat. Acad. Sci. USA, 58(6):2193-2195 (1967).

137. Sweeney, W. J., The D-Neumann problem, Acta Math., 120(3-4):223-277 (1968).

138. Tricomi, F. G., Solution of a problem of Goursat for a particular hyperbolic system of partial differential equations [in Italian], Rend. Ist. Lombardo Sci. Lett. Sci. Mat. Fis. Chim. Geol., 99(1):104-109 (1965).

139. Vaillant, J., Multiple characteristics and bicharacteristics of linear systems of partial differential equations with constant coefficients [in French], Ann. Inst. Fourier, 15(2):225-311 (1965).

140. Yamamoto, M., On Cauchy's problem for a linear system of partial differential equations of first order, Proc. Japan. Acad., 42(6):555-559 (1966).

141. Yamanaka, T., On the Cauchy problem for Kowalevskaja [Kovalevskaya] systems of partial differential equations, Comment. Math. Univ. St. Pauli, 15(2): 67-89 (1967).

142. Zerner, M., Singular solutions of partial differential equations [in French], Bull. Soc. Math. France, 91:302-226 (1963).

Fredholm Operators and Their Generalizations

S. N. Krachkovskii and A. S. Dikanskii

The aim of the present article is to review the main lines of development of the theory of the Φ-, Φ_+- and Φ_--operators primarily in the last ten years. A brief outline of earlier theories (Fredholm, Riesz-Schauder, Noether) affording the background for the Φ, Φ_+, Φ_--operators is given in § 1. In § 2 the basic results concerning the Φ, Φ_+, Φ_--operators in Banach spaces are presented, including the spectral and algebraic aspects of the theory. In § 3 the analogous problems in locally-convex and locally-bounded topological vector spaces are discussed. Finally, § 4 is given over to the abstract determinant theory of Fredholm. The problem area associated with the topological aspects of the theory (in the sense of K-theory [7], etc.) are not covered in the present review.

§ 1. Introduction

In 1902 and 1903 Fredholm published two papers [105, 106], in which he investigated the integral equations

$$x(s) - \lambda \int_a^b K(s, t)\, x(t)\, dt = y(s), \tag{1}$$

which have since come to be known as Fredholm integral equations of the second kind. In those papers the resolvent of Eq. (1) with continuous kernel K (s, t) is expressed as the quotient of two integral analytic functions of the complex parameter λ, which are obtained as the limit of the determinants used to describe the solution of an algebraic system of equations approximating Eq. (1).

Also, however, in the course of development of the theory alternative postulates emerged (the "Determinantenfreihe Sätze" of Hellinger and Toeplitz), which did not contain any explicit mention of the indicated Fredholm integral functions. These postulates may be stated as follows:

For given λ the integral equation (1) either has for every arbitrary continuous function $y(s)$ one and only one continuous solution $x(s)$, in particular the solution $x = 0$ for $y = 0$, or the corresponding homogeneous equation

$$x(s) = \lambda \int_a^b K(s, t)\, x(t)\, dt \tag{1_0}$$

has a finite number r of linearly-independent solutions $x_1, x_2, \ldots, x_r$. In the former case the associated equation corresponding to (1)

$$\varphi(s) - \lambda \int_a^b K(t, s)\, \varphi(t)\, dt = \psi(s), \tag{2}$$

also always has a uniquely-determined solution; in the latter case, on the other hand, the associated homogeneous equation

$$\varphi(s) = \lambda \int_a^b K(t, s)\, \varphi(t)\, dt \tag{2_0}$$

also has r linearly-independent solutions $\varphi_1, \varphi_2, \ldots, \varphi_r$, and the inhomogeneous equation (1) has a solution if and only if $y(s)$ satisfies the r conditions

$$\int_a^b y(s)\, \varphi_i(s)\, ds = 0 \quad (i = 1, 2, \ldots, r). \tag{3}$$

The solution of Eq. (1) in this case is determined correct to an arbitrary term of the form $c_1x_1 + c_2x_2 + \ldots + c_rx_r$.

Later the determinant-free part of the Fredholm theory was substantiated, an event which, on the one hand, facilitated its description and, on the other, paved the way for liberal generalizations. The results of Riesz [174] and the general notion introduced by Schauder [185] of the adjoint operator in a Banach space made it possible to formulate the following two postulates, analogous to the above Fredholm alternative postulates, for linear completely-continuous (compact) operators V acting in a Banach space X:

1. The homogeneous equation

$$Tx \equiv x - \lambda Vx = 0 \tag{4}$$

and its adjoint equation

$$T^*\varphi \equiv \varphi - \lambda V^*\varphi = 0 \tag{5}$$

have for given λ the same finite number of linearly-independent solutions $x_1, x_2, \dots, x_r$ and $\varphi_1, \varphi_2, \dots, \varphi_r$.

2. It is necessary and sufficient for the solvability of the inhomogeneous equation

$$Tx \equiv x - \lambda Vx = y, \tag{6}$$

that $\varphi_i(y) = 0$ $(i = 1, 2, \dots, r)$; if this condition holds and x_0 is some solution of Eq. (6), then its general solution has the form

$$x_0 + \sum_{i=1}^{r} c_i x_i,$$

where c are arbitrary constants. Analogously, it is necessary and sufficient for the solvability of the adjoint equation

$$T^*\varphi \equiv \varphi - \lambda V^*\varphi = \psi, \tag{7}$$

that $\psi(x_i) = 0$ $(i = 1, 2, \dots, r)$; if this condition holds and φ_0 is a solution of Eq. (7), then its general solution has the form $\varphi_0 + \sum_{i=1}^{r} c_i\varphi_i$.

Moreover, it has been verified that the set of values of λ for which Eqs. (4) and (5) are nontrivially solvable does not have finite accumulation points.

In [174] Riesz also investigated sequences of spaces of null elements of the operator $T_\lambda = 1 - \lambda V$:

$$N(T_\lambda) \subset N(T_\lambda^2) \subset \dots \tag{8}$$

and sequences of domains of values

$$R(T_\lambda) \supset R(T_\lambda^2) \supset \ldots . \tag{9}$$

Riesz proved that there exists a natural number m, beginning with which all spaces of sequence (8), as well as (9), coincide, the strict embedding holding for smaller index numbers. Also, for the indicated number m the space X admits the direct sum decomposition $X = N(T_\lambda^m) \oplus R(T_\lambda^m)$.

The next step in the development of the theory was the transition to a linear bounded operator T acting from a Banach space X into a Banach space Y and having the following properties:

1) The equations

$$Tx = 0 \tag{10}$$

and

$$T^*\varphi = 0 \tag{11}$$

have the same finite number of linearly-independent solutions [i.e., $\dim N(T) = \dim N(T^*)$].

2) The equation

$$Tx = y \tag{12}$$

is normally solvable, i.e., solvable if and only if $\varphi(y) = 0$ for any solution φ of Eq. (11). In this case the operator T is said to be normally solvable.

Nikol'skii [51] showed that each of the representations $T = U + K$ and $T = U + V$, where U is an invertible operator ($UX = Y$) and K, V are a finite-dimensional and completely-continuous operator ($X \to Y$), respectively, is necessary and sufficient in order for T to satisfy conditions 1) and 2).

If instead of the Fredholm integral equation (1) we consider the following singular integral equation with integral interpreted in the sense of the Cauchy principal value:

$$a(t)\varphi(t)+\frac{b(t)}{\pi i}\int_L\frac{\varphi(\tau)}{\tau-t}d\tau+\int_L K(t,\tau)\varphi(\tau)d\tau=f(t), \tag{13}$$

where the kernel $K(t, \tau)$ and the known functions a, b, f belong to certain Hölder classes and the solution φ is sought in one of those classes, then, as shown by Noether [151], postulates analogous to the Fredholm alternatives are valid, differing from the latter only insofar as the numbers of linearly-independent solutions of the homogeneous equation and its adjoint are, in general, distinct. An abstract analysis of an equation of the form (13) in normed rings has been carried out by Khalilov [62-64]. Atkinson [6] has investigated the normally-solvable equation (12), assuming that the numbers $n(T) = \dim N(T)$ and $d(T) = \dim N(T^*)$ are finite. He referred to the corresponding operator T as a generalized Fredholm operator. Atkinson set forth a necessary and sufficient condition generalizing the above condition of Nikol'skii. Specifically, in order for the operator $T(X \to Y)$ to be a generalized Fredholm operator, it is necessary and sufficient that there exist bounded operators $U, V(Y \to X)$ such that

$$UT = I_1 - L_1,\ TV = I_2 - L_2, \tag{14}$$

where I_1 and I_2 are identity operators and L_1 and L_2 are finite-dimensional (or completely-continuous) operators in X and Y, respectively. The bounded operators U and V subsequently came to be known as the left and right regularizers, respectively. For the given type of operators Atkinson, in the same paper, introduced the concept of the index $\varkappa(T) = n(T) - d(T)$ and proved that the index of the product of two generalized Fredholm operators is equal to the sum of the indices of each. He also verified the stability of those operators under completely-continuous and small (in the norm) perturbations. At almost the same time similar results were deduced by Gol'dman in his dissertation and by Gokhberg [23, 24]. In [14] M. A. Gol'dman proved that for the stability of the normal solvability of the operator T under small perturbations and also under arbitrary completely continuous perturbations, it is necessary and sufficient that at least one of the numbers $n(T)$ and $d(T)$ be finite.

Many papers were devoted to the properties of the operator $T_\lambda = I - \lambda A$, where A is a linear bounded operator. It was established that the set of values of λ for which T_λ is a generalized Fredholm operator is an open set, in every connected component of which the index $\varkappa(T_\lambda)$ is constant. A number of results were also obtained on the properties of the sequences of spaces (8) and (9) in those connected components.

All these and other results generalized to the case of a closed linear operator have been characterized in detail in the celebrated monograph of Gokhberg and Krein [30], which was published in 1957 and contains an exhaustive bibliography. For this reason we shall touch only superficially on those papers from the era preceding 1957 which are cited in [30].

§ 2. The Φ-Operators and Their Generalizations in Banach Spaces

1. A linear operator T acting from a linear topological space X into a linear topological space Y is said to be normally solvable if the equation

$$Tx = y \tag{15}$$

is solvable if and only if $\varphi(y) = 0$ $(y \in Y)$, where φ is any element of $R(T)^\perp \subset Y^*$, i.e., is such that $\varphi(R(T)) = 0$. If X and Y are locally-convex spaces and the domain of definition D(T) is dense in X, then there exists an adjoint operator T^*. In this case the definition of normal solvability given here coincides with the definition of § 1.

Normal solvability of T implies normal solvability of T^*, i.e., if T is normally solvable, then the equation

$$T^*\varphi = \psi \tag{16}$$

is solvable if and only if $\psi(x) = 0$ for any $x \in N(T)$. The converse is also true. Moreover, each of these two properties (normal solvability of the operator T and normal solvability of the operator T^*) is equivalent to closure of R(T), as well as closure of $R(T^*)$. The equivalence of the four enumerated properties for linear bounded operators in Banach spaces has been proved by Banach [76]. For bounded operators in a Frechet space this fact has been verified by Dieudonne and Schwartz [101]. For closed operators

T acting in Banach spaces Rothe and Markus (see the translator's note to the Russian edition of [81]) proved the equivalence of the closure of R (T) and R (T^*), and Kato [132] proved the equivalence of the normal solvability of T and T^*: if we denote by $\widetilde{T}$ the operator generated by the operator T (X → Y) in the factor space $\widetilde{X}$ = X/N (T), and by γ (T) the quantity $\|\widetilde{A}^{-1}\|^{-1}$, then, as shown by Kato [132], $\gamma(T) = \gamma(T^*)$, and T is normally solvable if and only if $\gamma(T) > 0$. The equivalence of all four conditions for closed operators in Frechet spaces has been proved by Browder [81]. In locally-convex spaces the normal solvability of T is equivalent to the closure of R (T) (see, e.g., [182]).

2. All the spaces discussed throughout the remainder of § 2, unless stated otherwise, are assumed to be Banach spaces. Let T be a closed linear operator acting from the space X into the space Y. We introduce the notation n (T) = dim N (T) and d (T) = dim R $(T)^{\perp}$ (= dim Y/R (T)).* If D (T) is dense in X, we have d (T) = n (T^*). A normally-solvable closed operator T with finite n (T) and d (T) is called a Φ-operator; if n (t) < +∞ and d (T) = ∞, T is called a Φ_+-operator; and if n (T) = ∞ and d (T) < +∞, it is called a Φ_--operator (see [30]).

If we interpreted d (T) as the dimension of the vector space algebraically complementary to R (T), then, as shown by Kato [132] (see also [109, 133]), the finiteness of d (T) implies normal solvability of the closed operator T. Consequently, the Φ- (and Φ_--) operators can be determined as closed operators with finite n(T) and d (T) [accordingly, with finite d (T), where d (T) is interpreted in the sense just indicated].

Note that the Φ-operators are referred to by a number of authors as Fredholm, Noether, or F-operators, and the Φ_+- and Φ_--operators as semi-Fredholm operators. The set of all Φ- (Φ_+-, Φ_--) operators acting from X into Y will be denoted henceforth by Φ (X, Y) [Φ_+(X, Y), Φ_-(X, Y), respectively].

We set in correspondence with every Φ-operator T an integer $\varkappa$ (T), called the index, according to the formula

$$\varkappa(T) = d(T) - n(T). \tag{17}$$

*The dimension of a subspace L of the Banach space $\mathfrak{B}$ is defined as the lowest power ω_L of sets whose linear hull is dense in L.

The index corresponding to the Φ_+-operator is equal to $-\omega_{N(T)}$, and the index corresponding to the Φ_--operator is equal to $\omega_{R(T)\perp}$.

Atkinson's results [6] on the existence of the right and left regularizer have been carried over to the Φ-operators by Aliev [3]. Mikhlin [49] has shown that if a closed operator $T(X \to Y)$ admits a left regularizer, it is a Φ_+-operator. Analogously, if it admits a right regularizer, it is a Φ_--operator (see [2]).

Let B_l ($l = 1, 2, 3$) be Banach spaces, where $B_2 \subset B_1$ and the embedding operator is completely continuous; the operator T is closed and acts from $D(T) \subset B_2$ into B_3. Then a necessary and sufficient condition in order for T to be a Φ_+-operator is fulfillment of the a priori estimate

$$\|u\|_2 \leqslant C(\|Tu\|_3 + \|u\|_1), \quad u \in D(T),$$

with the constant C independent of u (see, e.g., [2]).*

A generalization of this result to the closed T-completely-continuous mapping of B_2 into B_1 is given in Avantaggiati's paper [75].

Assuming the given linear operators are closed and densely defined, Schechter [186] has proved that:

1) if $A \in \Phi(X, Y)$, $E: Y \to W$, $EA \in \Phi(X, W)$, then $E \in \Phi(Y, W)$;

2) if $E \in \Phi(Y, W)$, $A: X \to Y$, $EA \in \Phi(X, W)$, then the restriction of A to $D(EA)$ is a Φ-operator acting from X into $D(E)$;

3) if $A: X \to Y$ and there exists a bounded linear operator E acting from Y into W such that $EA \in \Phi(X, W)$, then $A \in \Phi(X, Y)$.

In addition, Avantaggiati [75] has proved that:

4) if $EA \in \Phi_+(X, W)$, then $A \in \Phi_+(X, Y)$;
5) if $EA \in \Phi_+(X, W)$ and $\dim(D(E)/\overline{R(A)}) < \infty$, then $E \in \Phi_+(Y, W)$.

Gokhberg and Zambitskii [28, 29] have investigated normally-solvable operators in a Hilbert space H containing as a dense subset a Banach space B with norm satisfying the condition $\|x\|_B \geq C\|x\|_H$ ($x \in B$). In the ring $L(B)$ of all bounded operators acting into B they delineated a set Π of operators having the property that

*The analogous assertion (in terms of the adjoint operator) also holds for the Φ_--operator.

for every $A \in L(B)$ there exists an operator $A^+ \in L(B)$ such that $(Ax, y)_H = (x, A+y)$ $(x, y \in B)$. The operator $A \in \Pi$ is said to be B-normally solvable if the equation $Ax = y$ $(y \in B)$ is solvable if and only if $(y, \psi)_H = 0$ for any solution $\psi \in B$ of the equation $A^+\psi = 0$. This definition does not require the notion of the conjugate space. It is demonstrated that if $A (\in \Pi)$ and A^+ are Φ-operators, then they are B-normally solvable if and only if $\varkappa(A) = -\varkappa(A^+)$. Some generalizations to the case of unbounded operators have been given in Paraska's paper [54].

3. In this section we survey papers published after [30] and devoted to sets of points λ of a complex plane such that the closed operator $A_\lambda = A - \lambda I$ acting in X is a Φ-, Φ_+-, or Φ_--operator, along with other papers relating to the same problem area. The indicated sets are referred to, respectively, as Φ-, Φ_+-, or Φ_--sets. We know [30] that these sets are open and, therefore, decompose into connected components, where in every such component the numbers $n(A_\lambda)$ and $d(A_\lambda)$ are constant except for an isolated set of points.

Markus [44, 45] has investigated an operator function A_λ holomorphic in a simply-connected domain G of the complex plane of λ, the values of which are closed linear operators acting from X into Y. Let λ_0 be an arbitrary point in G, and let $|\lambda - \lambda_0| < \rho$ be the disk in which A_λ admits decomposition into the following series convergent on the norm of the operators:

$$A_\lambda = A_{\lambda_0} + \sum_{i=1}^{\infty} (\lambda - \lambda_0)^i C_i.$$

Let $x_0 \in N(A_{\lambda_0})$ $(x_0 \neq 0)$. Let us denote by $\rho(x_0, A_{\lambda_0})$ the largest of the nonnegative integers, for each of which there exist vectors $x_{\mu 0} = x_0, x_{\mu 1}, \ldots, x_{\mu\mu}$ such that

$$\sum_{i=0}^{k} C_i x_{\mu, k-i} = 0, \ (k = 1, \ldots, \mu), \ C_0 = A_{\lambda_0}.$$

If among these numbers there is not a largest one, we set $\rho(x_0, A_{\lambda_0}) = \infty$. The linear set consisting of all vectors $x \in N(A_{\lambda_0})$ for which $\rho(x, A_{\lambda_0}) = \infty$ is denoted by $\mathfrak{R}(A_{\lambda_0})$. In [45] it is proved that if A_λ is a Φ- (Φ_+- or Φ_--) operator for any $\lambda \in G$, there exists an isolated set $\Gamma \subset G$ such that for all $\lambda \in G \setminus \Gamma$ the function $n(A_{\lambda_0})$ preserves the same value n_0, whereas if $\lambda \in \Gamma$, then $n(A_\lambda) > n_0$. Also, for all $\lambda \in G$ the function $\bar{n}(A_\lambda)$ has the same value: $\bar{n}(A_\lambda) = n_0$.

In the special case when $X = Y$ and $A_\lambda = A - \lambda I = A_{\lambda_0} - (\lambda - \lambda_0)I$ these results signify that the number n_0 of independent infinite chains of null elements of the operator A_λ is identical for all $\lambda \in G$, while $n(A\ ) > n_0$ on the isolated set $\Gamma \subset G$ [38]. The so-called infinite chain of null elements of A_λ refers to a sequence of elements $x_1, x_2, x_3, \ldots$, such that $A_\lambda x_1 = 0$ and $A_\lambda x_n = X_{n-1}$ $(n = 2, 3, \ldots)$; several infinite chains are independent if the nulls of A_λ heading them are linearly independent of one another. At points $\lambda \in \Gamma$ there are, besides the infinite chains, finite chains $x_1', x_2', \ldots, x_p'$, which are characterized by the fact that $A_\lambda x_1' = 0$, $A_\lambda x_2' = x_1', \ldots, A_\lambda x_p' = x_{p-1}'$, whereas the equation $A_\lambda x = x_p'$ is unsolvable.

For the case $A_\lambda = A - \lambda I$ Gol'dman and Krachkovskii [18] have investigated the following problem. Let

$$\mathfrak{N}_\lambda = \bigcup_{n=1}^{\infty} N(A_\lambda^n), \quad \mathfrak{M}_\lambda = \bigcap_{n=1}^{\infty} R(A_\lambda^n),$$

where $\lambda \in F$ and F is a union of Φ-, Φ_+-, and Φ_--sets. It is proved that $\Gamma : \lambda \in F$, $\mathfrak{N} \cap \mathfrak{M}_\lambda \neq \mathfrak{M}_\lambda$, is an isolated set and that if G is some connected component of F, then on $G \setminus \Gamma$ the subspaces $\bar{\mathfrak{N}}_\lambda$ and $\mathfrak{M}_\lambda$ are constant. For $\mathfrak{M}_\lambda$ this result may also be found in papers by Gokhberg and Markus [30, 45]. The fact that $\bar{\mathfrak{N}}_\lambda$ is constant on $G \setminus \Gamma$ means that the space defined by all elements from which are formed infinite chains of zero elements of A_λ is constant along $G \setminus \Gamma$. These results are generalized as follows [19]. Let $\{B\}$ be a set of linear bounded operators in X that commute with A and with each other, and let $\mathfrak{B}$ be the closure of its linear hull; all operators of $\mathfrak{B}$ commute with A and with each other. In $\mathfrak{B}$ we delineate the set $\mathfrak{J}$, which consists of operators S such that $A_S = A - S$ is a Φ-, Φ_+-, or Φ_--operator, i.e., $R(A_S)$ is closed in X and at least one of the numbers $n(A_S)$ or $d(A_S)$ is finite. The set $\mathfrak{J}$ is open in $\mathfrak{B}$, and the set Γ of those $S \in \mathfrak{J}$ for which A_S has zeros not contained in $\mathfrak{M}_S = \bigcap_{n=1}^{\infty} R(A_S^n)$, is closed in $\mathfrak{J}$. If G is a connected component of the set F, then the spaces $\bar{\mathfrak{N}}_S = \overline{\bigcup_{n=1}^{\infty} N(A_S^n)}$ and $\mathfrak{M}_S$ are constant along $G \setminus \Gamma$.

The analogous problems for operator sheaves of the form

$A + \lambda B$, where A is a closed, and B an A-bounded [i.e., such that $D(A) \subset D(B)$, whence $\|Bx\| \leq \sigma \|x\| + \tau \|Ax\|$ for any $x \in D(A)$] linear operators from X into Y, have been treated in papers by Kato [132], Kaashoek [127], Oliver [153], and Förster [103, 104]; these authors generalize the established results concerning the zeros and domains of values of the iterated operators T^n $(n = 1, 2, \ldots)$ for the case X = Y to the case of different spaces.

Often, in addition to the numbers n(A) and d(A), two others are investigated: the ascent $\alpha(A)$ of the operator A, i.e., the smallest of the numbers k for which $N(A^k) = N(A^{k+1})$ [so that $N(A) \subset N(A^2) \subset \ldots \subset N(A^{\alpha}) = N(A^{\alpha+1}) = \ldots$], and the descent δ (A) of A, i.e., the smallest of the numbers k for which $R(A^k) = R(A^{k+1})$ [so that $R(A) \subset R(A^2) \subset \ldots \subset R(A^{\delta}) = R(A^{\delta+1}) = \ldots$]. In the case when such finite k do not exist the numbers $\alpha(A)$ and $\delta(A)$ are set equal to ∞. For the operators $V_\lambda = V - \lambda I$, where V is a completely-continuous operator, the equation $\alpha(V_\lambda) = \delta(V_\lambda)$ has been verified by Riesz [174]. Taylor [199] and Kaashoek [128] give different necessary and sufficient conditions for the numbers $\alpha(A)$, $\delta(A)$, n(A), and d(A) to be finite and analyze the relationships between them.

The Riesz domain $\mathfrak{R}(A)$ is defined as the set of λ for which the numbers $\alpha(A - \lambda I)$ and $\delta(A - \lambda I)$ are finite. Let F (A) denote the Φ-set of A (sometimes called the Fredholm domain of the operator A). Caradus [85] has proved that if A is densely defined, the set $F(A) \cap \mathfrak{R}(A)$ is open and on its connected components $\alpha(A - \lambda I)$ and $\delta(A - \lambda I)$ are equal. Kaashoek [129] has proved that the points of $\mathfrak{R}(A) \cap F(A)$ either belong to the resolvent of the set $\rho(A)$ of A or are poles of finite order of the resolvent of $(A - \lambda I)^{-1}$. If $\rho(A)$ is nonempty, then, as shown by Kasshoek and Lay [130], $\mathfrak{R}(A) \cap \sigma(A)$ is the set of poles of the resolvent $(A - \lambda I)^{-1}$. Kaashoek [129] has shown, in addition, that if the condition $\rho(A) = \emptyset$ is replaced by the condition that d(A) is dense in X, the foregoing assertion remains in effect. Some of these results have also been deduced by Schechter [186].

Ruston [179] introduced the concept of the Riesz operator. It is defined as a bounded operator $K: X \to X$ satisfying the following conditions:

1) $K_\lambda = K - \lambda I$ is a Φ-operator for all $\lambda \neq 0$;

2) $\alpha(K_\lambda) = \delta(K_\lambda) < \infty$;

3) The eigenvalues of the operator K can accumulate only at zero;

4) $\varkappa(K_\lambda^n) = 0$ for $n \geq 0$ and all $\lambda \neq 0$.

Caradus [84] has shown that, given proper additional restrictions on the operator K, conditions 1)-4) are independent. He has also shown (see also Pietsch [161] that the set of all Riesz operators is congruent with the set of all continuous operators T acting in X for which the Φ-set $F(T) = \{\lambda : |\lambda| > 0\}$. West [203] has demonstrated that in general the set of all Riesz operators does not form a closed subset in the algebra of all bounded operators. Gramsch [112] has proved (in a study of operators in locally-bounded spaces) the stability of the Φ-operator T and the stability of its index $\varkappa(T)$ under perturbation of T by a Riesz operator.

A generalization of the class of Riesz operators is found in the class of meromorphic operators introduced by Caradus [86], which consists of those bounded operators acting in X for which the nonzero points of the spectrum are poles of the resolvent. If it is required in addition that every nonzero eigenvalue of a meromorphic operator have finite multiplicity, this defines the subclass of Riesz operators. In [87] Caradus investigates the notion of the meromorphic operator on closed operators for which the nonzero and noninfinite points of the spectrum are poles of the resolvent.

In a number of papers the case is treated in which the Φ-set of the operator A_λ spans the entire complex plane. In particular, it has been proved in papers by Kaniel and Schechter [131] and Kaashoek and Lay [130] that $(\mu - T)^{-1}$ is a Riesz operator for any $\mu \in \rho(T)$ and, conversely, if $(\mu - T)^{-1}$ is a Riesz operator for some $\mu \in \rho(T)$, then F (T) spans the entire complex plane. Caradus [88] has generalized the familiar theorem of Gokhberg and Krein [30] stating that if T is a linear bounded operator acting in the Banach space X and F (T) coincides with the entire complex plane, then X is finite-dimensional. Specifically, he proved that if $D(T) = X$ and $d(T - \lambda I) < \infty$ [or $n(T - \lambda I) < \infty$] and $R(T - \lambda I)$ is closed for any λ, then X is finite-dimensional. The result remains valid if the above conditions are replaced by the following: D (T) is closed, $n(T - \lambda I) < \infty$ for all λ, $d(T - \lambda_0 I) < \infty$ for some λ_0, and $R(T - \lambda I)$ is closed for all λ. Kaashoek and Lay [130] have obtained a similar result for closed operators.

4. In this section we consider the problems associated with the stability of the Φ, Φ_+, Φ_--operators under perturbation by operators which are to be considered small in some sense.

It is well known (see [53]) that on the set C of all closed operators acting from X into Y it is possible to introduce a metric $\rho(A, B) = \theta(G_A, G_B)$, where $\theta(L_1, L_2)$ is the angle spanned by the two subspaces L_1 and L_2, and G_A and G_B are the graphs of the operators $A, B \in C$ in $X \times Y$. Paraska [53] has proved that if $A \in C$ is a Φ -operator (Φ_+-operator), there exists a $\delta > 0$ such that if $B \in C$ satisfies the inequality $\rho(A, B) < \delta$, then B is a Φ_+-operator (Φ_--operator), $n(B) \leq n(A)$ $[d(B) \leq d(A)]$, and $d(B) = d(A)$ $[n(B) = n(A)]$. But if A is a Φ-operator and $\rho(A, B) < 0$, then B is also a Φ-operator, $\varkappa(A) = \varkappa(B)$, $n(B) \leq n(A)$, and $d(B) \leq d(A)$. The same results have been obtained by Neubauer [148], in particular for many-valued mappings. Earlier Cordes and Labrousse [94] proved the same results for the case of closed operators acting in a Hilbert space H, demonstrating that for bounded operators the topologies defined by means of the matrics $\rho(A, B)$ and $\|A - B\|$ are equivalent. They also proved a result bearing on the complete system of homotopic invariants of the Φ-, Φ_+-, and Φ_--operators, namely that a closed operator A is homotopic to a closed operator B (i.e., there exists a continuous family of operators A_t, $0 \leq t \leq 1$, such that $A_0 = A$ and $A_1 = B$) if and only if $\varkappa(A) = \varkappa(B)$. This result is equivalent to the statement that an open set of Φ-, Φ_+-, and Φ_--operators can be partitioned in C into connected components, in each of which the operator index has a constant value, which differs in different components. For bounded operators acting in a Hilbert space the second part of the foregoing result was deduced by Gokhberg, Markus, and Fel'dman [33] (for more general results see Kuiper [138] (also [7]) and Shvarts [66]). Neubauer [148] has proved the same result for Φ-operators acting from a Banach space X into a separable Banach space Y (see also [7*]). For a Hilbert space H Cordes and Labrousse [94] have verified that, given arbitrary $\varkappa_0$, for some $|\varkappa_0| \leq \dim H$ there exists a Φ-, Φ_+-, or Φ_--operator A such that $\varkappa(A) = \varkappa_0$.

Coburn and Lebow [90, 91] have proved for bounded (closed) operators acting in a separable Hilbert space that the closure of the set of Φ-operators of index k is the complement of the set of all Φ, Φ_+, Φ_--operators with index distinct from k, i.e., if A is not a Φ-, Φ_+-, or Φ_--operator, then every neighborhood of A contains Φ-operators with any index. Gramsch [112] has proved the following for complete locally-bounded spaces: Let there be given on the set of all Φ-operators acting in the space X an integer-valued function $\nu(T)$ with the following properties: 1) $\nu(ST) = \nu(S) + \nu(T)$; 2) $\nu(S + F) = \nu(S)$, where F is a finite-dimensional operator in X; 3) $\nu(T) = 0$ if T is invertible. Then there exists a rational number r such that

$$\nu(T) = r\varkappa(T)$$

for all Φ-operators acting in X, where $\varkappa(T)$ is the index of T.

Condition 3), as shown by Gramsch [112], can be replaced by the requirement of connectivity on the part of the group of invertible operators. Without condition 3) the above result has been obtained for Hilbert spaces by Gokhberg, Markus, and Fel'dman [33].

5. We now consider certain classes of operators endowed with the property that every operator of the given class, on perturbing any Φ- (Φ_+- or Φ_--) operator, preserves its properties. We denote by L (X, Y) the Banach algebra of all bounded linear operators acting from X into Y. An operator $V \in L(X, Y)$ is said to be a Φ-admissible (Φ_+- or Φ_--admissible) perturbation if for any Φ-operator (Φ_+- or Φ_--operator) $A \in L(X, Y)$ the operator A + V is a Φ-operator (Φ_+- or Φ_--operator) (see [32]). It has long been known (see [30]) that finite-dimensional and completely-continuous operators are Φ-, Φ_+-, and Φ_--admissible perturbations. Kato [132] (see also [133]) introduced a class of operators which are Φ- and Φ_+-admissible perturbations. These are operators (which he calls strictly-singular operators) $A: X \to Y$ whose restriction to any infinite-dimensional subspace in X is not an isomorphic embedding. Kato [132] has shown that the strictly-singular operators form a closed subspace in L (X, Y) which is a two-sided ideal when X = Y. Gokhberg, Markus, and Fel'dman [33] have shown that completely-continuous operators do not exhaust the set of all strictly-singular operators and that the latter are not, in general, Φ_--admissible perburbations. The same results have been deduced by Goldberg and Thorp [110]. Pelczyński [154] introduced the concept of strictly-cosingular operators, which represents the dual of the concept of strictly-singular operators: An operator $A \in L(X, Y)$ is said to be strictly cosingular if there does not exist an infinite-dimensional

Banach space E with superpositions $h_1 : X \to E$ and $h_2 : Y \to E$ such that $h_2 T = h_1$. Vladimirskii [8, 10] has shown that strictly-cosingular operators are Φ_--admissible perturbations but are not, in general, Φ_+-admissible perturbations. The set of all strictly-cosingular operators is closed in L (X, Y) and forms a two-sided ideal when X = Y. Gokhberg, Markus, and Fel'dman [33] have proved that the largest two-sided ideal R contained in the set of Riesz operators is closed, that the algebra L (X)/R is semisimple, and that the following chain of embeddings of the two-sided ideals of Φ-admissible perturbations holds:

$$\mathfrak{F}(X) \subsetneq \overline{\mathfrak{F}}(X) \subset \mathfrak{K}(X) \subsetneq \mathfrak{S}(X) \subsetneq R,$$

where $\mathfrak{F}(X)$, $\mathfrak{K}(X)$, $\mathfrak{S}(X)$ denote the two-sided ideals of finite-dimensional, compact, and strictly-singular operators, respectively. Strictly-singular operators are also discussed in papers by Kleinecke [134] and Whitley [204].

Neubauer [149] has verified for closed many-valued mappings from X into Y the analogs of theorems on the perturbation of Φ-, Φ_+-, Φ_--operators by completely-continuous operators.

6. For the conclusion of § 2 we cite some results associated with the algebraization of the theory.

Khalilov [62, 64] has translated the fundamental theorem of Noether [151] concerning the singular integral equations (13) to the case of abstract singular equations in a normed ring R:

$$ux + vS(x) + T(x) = y,$$

where u, v, and y are given elements of R, x is an unknown element of R, S is a linear singular operator (i.e., one such that $S^2 = I$ and for any $x \in R$ the operator $xS - Sx$ is a Φ-operator with zero index), and T is a regular operator. Khalilov's results have been generalized by Agaev [1] and Cherskii [65] to equations in a Banach space. Samko [58] has obtained a generalization of Cherskii's results, which permits all the concrete types of equations investigated thus far with Cauchy kernels, including equations with Cauchy kernels on an open contour and with discontinuous coefficients, to be incorporated into the general theory of abstract singular equations.

In a series of papers [164-169] Przeworska-Rolewicz has investigated equations with algebraic operators in linear spaces, generalizing the equation analyzed by Khalil, and for those equations proved theorems analogous to the theorems of Noether.

Yood [206] obtained new proofs of the theorems of Atkinson [6] and Gokhberg [23, 24] in terms of Banach algebras and arrived at a number of new results. In particular, he proved that if J is any nonzero two-sided closed ideal of the algebra L(X) of all bounded linear operators acting in a Banach space X and is contained in the set of all Φ-admissible perturbations, then in order for an operator $A \in L(X)$ to be a Φ-operator $(X \to X)$ it is necessary and sufficient that the class of residues $\tilde{A}$ in the factor algebra $\tilde{L} = L/J$ be invertible in $\tilde{L}$. Yood's results were generalized by Gokhberg, Markus, and Fel'dman [33] (see § 5) and for rings without topology by Gramsch [113].

In the algebra L(X) Brauer and Cordes [80] have investigated the operators $T = I - B$ (which they called quasi-Fredholm operators), where B is a Φ-operator, and proved that T is a quasi-Fredholm operator if and only if the class of residues of the factor ring L(X) on the ideal of completely-continuous operators containing the element T is quasiregular.

Cordes [92, 93] introduced the following generalization of the Φ-operators, which is associated with a C*-algebra. Let $\mathfrak{A}$ be a C*-algebra with unit element, and let J be a proper closed two-sided *-ideal. Then $A \in \mathfrak{A}$ is said to be a J-Fredholm operator if its factorization on J is an operator $\breve{A}$ admitting inversion in $\mathfrak{A}/J$. The case is analyzed when the ideal J is an R-algebra (see, e.g., [92] for a definition of the R-algebra). In particular, the ideal of compact operators is an R-algebra. Many of the results pertaining to the Φ-operators are valid for the J-Fredholm operators, but closure of the domain of values cannot take place. J-Fredholm operators with a closed domain of values are considered.

We note, finally, that the Φ-operators have been investigated in linear space without topology by Krupnik [42], Przeworska-Rolewicz and Rolewicz [171-173], Gol'dman and Krachkovskii [20], and Gramsch [113].

§ 3. The Φ-Operators and Their Generalizations in Topological Vector Spaces

The main part of the theory of Riesz-Schauder equations with completely-continuous (compact) operators in Banach spaces was adapted by Leray [142] to the case when the operators act in locally-convex spaces. Another method of carrying over the Riesz-Schauder theory to locally-convex spaces has been proposed by Altman [68, 69, 4], who, defining continuity in terms of sequences rather than neighborhoods, proved that an operator adjoint to a completely-continuous operator is completely continuous. This enabled him to generalize the known results for Banach spaces on the solvability of adjoint equations to locally-convex spaces. Using the results of Leray [142], Williamson [205] proved that for the equation

$$(I - V)\, x = y$$

with a completely-continuous operator V in a topological vector space E the Fredholm alternative is applicable, namely that either the operator $I - V$ maps E itself homeomorphically or there exists an element $x \in E$ $(x \neq 0)$ such that $(I - V)\, x = 0$; in this case the numbers of linearly-independent solutions of the homogeneous equation and its adjoint are identical.

We note that for arbitrary locally-convex spaces the complete continuity of an operator does not necessarily imply complete continuity of its adjoint, nor vice versa (see [135, 175, 56]). However, for all functional spaces used in analysis both facts are true, as shown by Raikov [57].

A generalization of the Riesz-Schauder theory to locally-convex spaces was also obtained by Ringrose. Unlike Leray [142] and Williamson [205], however, he does not presume compactness of the adjoint operator. Unlike Altman, Ringrose offers the concept of continuity by means of neighborhoods, so that the concept of the adjoint operator and adjoint space takes on the customary significance. Hukuhara and Sibuya [123-125] have also investigated the Riesz-Schauder theory in locally-convex spaces.

Shirota [188] has generalized to locally-convex spaces the result of Nikol'skii [51] stating that if for some degree $n \geq 1$ the map T^n of a Banach space X into itself is completely continuous, then the Riesz-Schauder theory applies to T and T^*.

Schaefer [182] introduced the concept of the σ-mapping. A continuous mapping T of a locally-convex space E into a locally-convex space F is called a σ-mapping if T satisfies the properties of the Φ-operator and is a topological homomorphism. In accordance with the preceding notation, we shall refer to such a mapping as a Φ-operator. The Φ_+- and Φ_--operators in locally-convex spaces are defined analogously (see [9, 46]). Schaefer [182] has proved that in order for a continuous mapping $T(E \rightarrow F)$ to be a Φ-operator it is necessary and sufficient that there exist continuous mappings U, $V(F \rightarrow E)$ such that representation (14) holds. We observe that in the proof of this assertion the openness of T is relied upon, thus implying the continuity of T^{-1} if T^{-1} exists. In complete metrizable topological vector spaces (not necessarily locally-convex) this fact ensues from the Banach open mapping theorem, a result that stems from the fact that in such spaces it is necessary to demand openness of the mapping in defining the Φ-operator.

For Φ-operators acting in locally-convex spaces Schaefer (182] has proved that the index of the composition of two such operators is equal to the sum of the index of each and that the Φ-operator and its index are stable under perturbation by compact mappings. The latter result was proved earlier by Schwartz [187] and Köthe [135]; the same result was also deduced for Φ_+-operators by Köthe [135]. For complete metrizable topological spaces Gramsch [112] obtained the same results for Φ-operators.

Separate results concerning the properties of Φ, $\Phi_\pm$-operators in locally-convex spaces, generalizing the known results for those operators in Banach spaces, were obtained by Audin [73, 74], Deprit [96-100], and Pietsch [161]. It was proved that the mappings adjoint to Φ, $\Phi_\pm$-mappings are mappings of the same type (Schaefer [183]; Pietsch [161]); the converse has been proved only for Φ-mappings acting in a cylindrical space (Pietsch [161]).

In [159] Pietsch shows that if T is a Φ_+-operator $(E \rightarrow F)$ and K is a compact mapping, then $n(T-\lambda K) = n \leq n(T)$, whence it follows that for the mapping $T - \lambda K$ there exists a larger countable set of points λ_ν of the complex plane that does not have a finite accumulation point and is such that $n(T - \lambda K) = n$ for $\lambda \neq \lambda_\nu$ and $n(T - \lambda_\nu K) > n$. The corresponding result has also been proved for Φ_--operators.

Markus and Fel'dman [46] have investigated the perturbation by bounded mappings of Φ-operators acting from a complete locally-convex space E into a locally-convex space F. Let V be a neighborhood of zero in a complete locally-convex space E, and let S(V, F) be the locally-convex space of all linear mappings bounded on V and acting from E into a locally-convex space F (i.e., such that the image of V is a bounded set in F), where S is assigned uniform convergence topology on V. Then S(V, F) is a topological ring with continuous quasi-inverse. The stability of the Φ-operator under perturbation by small mappings A from S(V, F) [i.e., such that there exists a neighborhood of zero $U \subset F$ such that $A(V) \subset U$], as well as by mappings $B_\lambda \in S(V, F)$, where B_λ is an analytic function of the complex variable λ, has been proved. Vladimirskii [9] has generalized these results to Φ_+-operators and in [9, 1*] has considered certain classes of $\Phi_\pm$-admissible perturbations in locally-convex spaces without assuming that the latter are complete.

Pietsch [161] has shown that the set of all Φ-operators acting in a locally-convex space E is open for definite subalgebras of the algebra of continuous endomorphisms.

In [160] Pietsch discusses linear mappings with Φ-operator properties, defined on any subsets of a locally-convex space E and acting in a locally-convex space F. He calls them φ-mappings and shows that the φ-mappings are closed. Analogous questions for topological vector spaces are examined in a paper by Gol'dman and Krachkovskii [22]. A number of propositions applicable to the Φ-operators are generalized to the φ-mappings. In particular, representations (14) [the first of (14) only on the domain of definition of the operator] constitute a necessary and sufficient condition for T to be a φ-mapping. However, with regard to the index of the composition of two φ-mappings (also a φ-mapping), if they are not defined on dense subsets, all that can now be stated is that the index is not less than the sum of the indices of each mapping. Tôgô and Shiraishi [200] have shown that the relation $D(T_1) + R(T_2) = F$ is a necessary and sufficient condition in order to have $\varkappa(T_1 T_2) = \varkappa(T_1) + \varkappa(T_2)$ (T_2: $E \to F$, T_1: $F \to G$).

In [160] Pietsch introduces the concept of the γ-mapping, which is applicable to differential operators. A mapping T acting from a locally-convex space E into a locally-convex space F is called a γ-mapping if there exists a compact subset $K \subset E$ such that T(K) is a neighborhood of zero in R(T) and the properties of the

Φ-operator are fulfilled. It is proved that the γ-mapping is a φ-mapping and that the γ-mapping is stable under perturbation by any linear continuous mappings. Theorems are also proved on compositions of γ-mappings and φ-mappings.

Presdorf [55] has proved that in a countably-normed space a necessary and sufficient condition in order for an operator A to be a Φ -operator is the existence of a continuous operator B such that BA is a Φ -operator.

In [183] Schaefer considers in a locally-convex space E weakly-continuous Φ-operators with index equal to zero, which he calls Fredholm mappings, and he investigates Riesz mappings. A Riesz mapping (not to be confused with the Riesz operator, § 2, part 3) is defined as a weak homomorphism T of the space E into itself such that $N(T^n)$ is the same finite-dimensional subspace for $n \geq n_0$ and $T^m(E)$ is the same closed subspace for $m \geq m_0$. It is proved that then $m_0 = n_0$ and the Riesz mapping is a Fredholm mapping if and only if $\bigcup_{n=1}^{\infty} N(T^n)$ is finite-dimensional. A point λ is called a Fredholm (Riesz) point if $T_\lambda = T - \lambda I$ is a Fredholm (Riesz) mapping.

Leray [142] and Altman [68] have proved that for a compact operator T the spectrum σ (T) is a closed bounded set. Schaefer [183] has proved that if T is a bounded linear mapping acting in a sequentially-complete locally-convex space E, then its spectrum σ (T) is a closed bounded set, and the set of Fredholm points is an open set decomposable into two connected components φ_1 (T) and φ_2 (T) such that φ_1 (T) consists of all Riesz points of the mapping T; hence $\varphi_2(T) \subset \sigma(T)$, and the points of the set $\varphi_1(T) \cap \sigma(T)$ are isolated. This result has been generalized to Φ- and Φ_+-operators by Markus and Fel'dman [46, 47]. Riesz mappings and their generalizations to the case when one of the numbers n_0 or m_0 is infinite have been investigated by Deprit [99].

Gramsch [112], relying on [111], has investigated the Φ-operators in complete locally-bounded spaces. It is well known that it is possible to introduce into a locally-bounded space a p-norm ($0 < p \leq 1$) characterized by the following properties: 1) $\|x\|_p > 0$ $(x \neq 0)$; 2) $\|\lambda x\|_p = |\lambda|^p \|x\|_p$; 3) $\|x + y\|_p \leq \|x\|_p + \|y\|_p$. The algebra $L(X)$ of all continuous endomorphisms of a complete locally-bounded space

X forms a complete p-algebra, where $\|ST\|_p \leq \|S\|_p \|T\|_p$, with S, T $\in$ L (X). Gramsch [112] has proved that the semigroup of all Φ-operators acting in a complete locally-bounded space is open in L (X) and decomposes into connected components, in each of which the index is constant. Other properties valid for the Φ-operators in Banach spaces also turn out to be valid in locally-bounded spaces. The properties of the Riesz operators are also generalized to those spaces (§ 2, part 3).

Przeworska-Rolewicz and Rolewicz [170] give an example illustrating that the theorem of Gokhberg and Krein [30] stating that if the Φ-set of a continuous operator T acting in a Banach space X coincides with the entire plane, then X is finite-dimensional, does not extend to more general spaces without the assumption of local boundedness. However, Markus and Fel'dman [47] have proved that if T is a linear bounded operator acting in a complete locally-convex space X and its Φ_+- or Φ_--set coincides with the entire complex plane, then X is finite-dimensional. It is also proved in [170] that in a complete locally-bounded space the functions n (T − λI) and d (T − λI) are constant on every connected component of the Φ-set, with the exception of a countable set of points at which the functions are semicontinuous above. For locally-convex spaces this is not true in general, but if the operator T is bounded, the statement is true for complete locally-convex spaces [47].

§ 4. Abstract Development of the Determinant Theory of Fredholm

After the classical papers of Fredholm [105, 106], in which the resolvent of an integral equation was investigated in the form of the quotient of two integer-valued functions of a complex variable, the subsequent elaboration of the theory in this particular area, involving the transition from the integral to an operator equation, was almost forty years in coming.

Thus, in 1941 Smithies [198] introduced the notion of the determinant for the operator I + T in an abstract Hilbert space, where I is the identity operator and T is the operator generated by a matrix (T_{ij}) possessing the property that $\sum_{i,j} |\tau_{ij}|^2 < \infty$. This notion coincides with the one studied by Carleman [89] for integral equations in L_2-space (Carleman proved that the numerator and denominator of the Fredholm resolvent are integer-valued functions, given the only restricting assumption that the kernel of the integral equation is square-summable on the plane; a new version of the proof of this result was proposed by Mikhlin [48]). An outline of the theory of the Carleman determinants and their minors is given in articles by Sikorski [193, 196].

The further development of the abstract theory of the Fredholm determinants began in the work of Ruston [176-179] and Grothendieck [119]. The notion investigated by them of the determinant and its minors falls in the class of nuclear operators (or operators with traces) in a Banach space. This class is narrower than the class of completely-continuous (compact) operators, but contains all operators of finite rank. The characteristic feature of the Fredholmian theories of Ruston and Grothendieck is their use of the apparatus of the tensor and outer products and t-normalization of the tensor products of Banach spaces, developed by Schatten [184] and Grothendieck [116-118], which enable one to delineate the required class of operators with traces. We now briefly describe this construction.

Let X be a Banach space, let X' be its conjugate, let x, y,... be elements of X, and let f, g,... be elements of X'. The tensor product $X \otimes X'$ is the set consisting of elements of the form $\sum_{i=1}^{n} x_i \otimes f_i$, which are identified with operators of finite rank acting in X: $\left(\sum_{i=1}^{n} x_i \otimes f_i\right) y = \sum_{i=1}^{n} f_i(y) x_i$. Consequently two elements $\sum_{i=1}^{n} x_i \otimes f_i$ and $\sum_{i=1}^{m} y_i \otimes g_i$ of $X \oplus X'$ are considered to be equivalent if they correspond to the same operator. Following Schatten [184], we equip $X' \otimes X'$ with the so-called γ-norm: $\gamma\left(\sum_{i=1}^{n} x_i \otimes f_i\right) = \inf \sum_{i=1}^{m} \| y_i \| \| g_i \|$, where the symbol inf is taken over all equivalent expressions of the given element, i.e., over all expressions $\sum_{i=1}^{m} y_i \otimes g_i$, for which,

$$\left(\sum_{i=1}^{m} y_i \otimes g_i\right) y = \sum_{i=1}^{n} f_i(y)\, x_i,$$

for any $y \in X$. The complete tensor product of spaces X and X' is defined as the completion $X \hat{\otimes} X'$ of the space $X \otimes X'$ on the γ-norm. This completion implies, in addition, expansion of the class of operators of finite rank by the adjunction of those operators which are the limits on the γ-norm of sequences of operators of finite rank. The resulting class of operators, called nuclear operators or operators with trace, is narrower than the class of operators than can be approximated with any error on the conventional norm by means of operators of finite rank (and is thus narrower than the class of compact operators). The indicated completion process uniquely determines for the nuclear operators a t-norm (as the limit of γ-norms), by means of which the concept of the trace of a nuclear operator $\hat{A}$ is obtained by passage to the limit as applicable to the traces $\operatorname{tr}(A) = \sum_{i=1}^{n} f_i(x_i)$ of operators of finite rank $A = \sum_{i=1}^{n} x_i \otimes f_i$ that are γ-convergent to $\hat{A}$.

Once the trace concept has been defined for all nuclear operators, the system of determinants is constructed as in the classical theory of the Fredholm integral equations. Thus, for example, the "denominator" and "numerator" of Fredholm are integer-valued analytic functions of the form

$$d(\lambda) = \sum_{n=0}^{\infty} d_n \lambda^n, \quad D(\lambda) = \sum_{n=0}^{\infty} D_n \lambda^n,$$

where $d_0 = 1$, $d_n = \frac{(-1)^n P_n}{n!}$, and

$$D_n = \frac{(-1)^n}{n!} \begin{vmatrix} I & n\,0 \ldots 0 \\ \hat{A} & \\ \hat{A}^2 & P_n \\ \hat{A}^n & \end{vmatrix}; \qquad P_n = \begin{vmatrix} \sigma_1 & n-1 & 0 \ldots 0\,0 \\ \sigma_2 & \sigma_1 & n-2 \ldots 0\,0 \\ \cdot & \cdot & \cdots \\ \sigma_{n-2} & \sigma_{n-3} & \sigma_{n-1} \ldots 2\,0 \\ \sigma_{n-1} & \sigma_{n-2} & \sigma_{n-3} \ldots \sigma_1\,1 \\ \sigma_n & \sigma_{n-1} & \sigma_{n-2} \ldots \sigma_2\sigma_1 \end{vmatrix}; \qquad \sigma_k = \operatorname{tr}(\hat{A}^k).$$

Bearing on a similar course in the development of the Fredholm theory is a paper by Saphar [181], in which the theory is formulated for operators in a Banach space such that their power is a nuclear operator.

A systematic and original conception of the theory of nuclear operators (and the theory of symmetrically-normed ideals of the algebra of linear bounded operators in general) in a Hilbert space, as well as the construction of their corresponding Fredholm determinants (based on subtle properties of the theory of analytic functions) are presented in the book of Gokhberg and Krein [32]. Their presentation is based on an investigation of the classes $\mathfrak{S}_p$ (p = 1, 2) of completely-continuous operators ($\mathfrak{S}_1$ and $\mathfrak{S}_2$ refer, respectively, to the class of nuclear operators and the class of Hilbert-Schmidt operators) characterized by a definite rate of decay of the eigenvalues (in the case of integral operators a detailed investigation of the decay rate of the eigenvalues has been given by Gel'fond [11]). For every operator $T \in \mathfrak{S}_p$ it proves possible to construct an infinite product describing a certain integer-valued function, i.e., the regularized characteristic determinant (whose zeros are the inverses of nonvanishing eigenvalues of the operator, where the multiplicity of the zero is equal to the number of linearly-independent null elements corresponding to the given eigenvalue), which act the part of the Fredholm determinant and thus incorporate the operators of classes $\mathfrak{S}_p$ (p = 1, 2) into the abstract Fredholm theory. We note, finally, that Ruston [178] defined a class of asymptotically-quasicompact operators (encompassing all compact operators), for which the representation of the resolvent as the quotient of two integer-valued functions and the corresponding Fredholmian formulas remain in effect. However, the construction of the Fredholm determinant (the regularized characteristic determinant) in this case is more complex.

Another avenue of development of the Fredholm theory was initiated by Leżański [143, 144] on the basis of the theory of multilinear functionals and did not rely on the direct products of Banach spaces or the traces of an operator. His theory of determinants for linear equations in Banach spaces was then expounded by Sikorski [189] in a somewhat modified form corresponding to more symmetric assumptions, by which it would be possible to deduce complete symmetry of the results for a given linear equation and its adjoint. Later Sikorski showed [190] that the nuclear operators always satisfy the postulates of Leżański, but the converse is not true in general Banach space, although it does hold for a Hilbert spacc and for reflexive Banach spaces. In the latter cases the approaches of Ruston, Grothendieck and Sikorski coincide [190]. In his review article on the theory of determinants in a Banach space and in a number of subsequent papers [195, 197] Sikorski characterizes a class of quasinuclear operators in a Banach space for which it is possible to formulate the Leżański determinant theory. This class may be concisely portrayed as follows. Let X be a Banach space, x, y,... its elements, f, g,... the elements of the conjugate space X', and E the Banach algebra of all linear bounded operators acting in X; an operator $T \in E$ is said to be quasinuclear if there exists a linear bounded functional F on E such that $fTx = F(xf)$ for any $x \in X$, $f \in X'$ [the symbol xf denotes the one-dimensional operator: $Ay = xf(y)$]. Then F is called the quasinucleus of the operator T, and the system of determinants (the Fredholm determinant and its minors) for the operator I + T is constructed on any quasinucleus F of T as follows:

On the Cartesian product $X'^m \times X^m$ consider a 2m-linear functional of the form

$$\theta_m\begin{pmatrix} f_1, \ldots, f_m \\ x_1, \ldots, x_m \end{pmatrix} = \begin{vmatrix} f_1x_1 \ldots f_1x_m \\ \cdots\cdots\cdots \\ f_mx_1 \ldots f_mx_m \end{vmatrix}.$$

If all the variables except the pair f_1, x_i are fixed, this a bilinear functional with respect f_1, x_i, which determines a certain operator from E. Let us denote by $F_{f_i x_i}$ the action of the functional F on that operator. Then the expression

$$D_{nm}(F) = F_{f_{n+1}x_{n+1}} \ldots F_{f_{n+m}x_{n+m}} \theta_{n+m}\begin{pmatrix} f_1, \ldots, f_{n+m} \\ x_1, \ldots, x_{n+m} \end{pmatrix}$$

is a 2n-linear functional, and

$$D_n\begin{pmatrix} f_1, \ldots, f_n \\ x_1, \ldots, x_n \end{pmatrix} = \sum_{m=0}^{\infty} \frac{1}{m!} D_{nm}(F), \quad n = 0, 1, 2, \ldots,$$

is the system of Leżański determinants for I + T. The corresponding system of Fredholm determinants becomes

$$D_n^*\begin{pmatrix} f_1, \ldots, f_n \\ x_1, \ldots, x_n \end{pmatrix} = D_n\begin{pmatrix} f_1, \ldots, f_n \\ Tx_1, \ldots, Tx_n \end{pmatrix}, \quad n = 0, 1, 2, \ldots.$$

Note that the canonical mapping $F \to T$ is not, in general, one-to-one, so that the construction of a system of determinants is not, in general, uniquely determined by the operator T. In many specific cases, however, it is known that the mapping of a class of nuclei onto its corresponding class of nuclear operators is one-to-one. But in the general case and for nuclear operators this problem remains unsolved (it is equivalent to the well-known problem of the approximation of a completely-continuous operator by operators of finite rank).

If T is a quasinuclear operator, then each of the endomorphisms y = Tx and g = T'f maps bounded sets into weakly-compact sets and weakly-compact sets into compact sets [117].

Moreover, for a quasinuclear operator T all powers T^n $(n > 1)$ are nuclear and, hence, compact. Also, if T_1 and T_2 are quasinuclear, T_1T_2 is nuclear [117].

A certain generalization of the Leżański-Sikorski theory has been given by Buraczewski [82] in an investigation in general linear spaces of the problem of constructing a system of determinants for linear operators having the properties of Φ-operators.

In [163] Pietsch introduced the concept of related endomorphisms for two separable locally-convex spaces E and G (endomorphisms T and $\mathfrak{T}$ are related in E and G, respectively, if there exist continuous linear mappings P: E → E and Q: G → E such that T = QP and $\mathfrak{T}$ = PQ) and showed that for two related endomorphisms T and $\mathfrak{T}$ the results of the Fredholm theory are transferable from one to the other. In particular, the existence in the Sikorski sense of a system of determinants for I + $\mathfrak{T}$ implies the existence of such a system for I + T. A special type of endomorphism (semi-integral endomorphism), which turns out to be related to a certain Hilbert-Schmidt endomorphism and has a Sikorski system of determinants, is given as an example.

Literature Cited

1. Agaev, G. N., On the theory of a singular equation in a Banach space, Trudy Inst. Fiz. Mat. Akad. Nauk Azerbaidzhan. SSR, 8: 23-37 (1969).
2. Agranovich, M. S., Elliptic singular integrodifferential operators, Usp. Mat. Nauk, 20(5): 3-120 (1965).
3. Aliev, F. S., Generalization of a theorem of S. M. Nikol'skii to the case of a closed operator, Dokl. Akad. Nauk Azerbaidzhan. SSR, 16(1): 7-11 (1960).
4. Altman, M., On the Riesz-Schauder theory of linear operator equations in spaces of type (B_0), Studia Math., 15(2): 136-143 (1956).

5. Andrunakievich, V. A. and Gokhberg, I. Ts., Linear equations in linear infinite-dimensional spaces, Uch. Zap. Kishinev. Univ., 5 : 63-67 (1952).
6. Atkinson, F. V., Normal solvability of linear equations in normed spaces, Mat. Sb., 28(1) : 3-14 (1951).
7. Atiyah, M., K-Theory; Lectures, Benjamin, New York (1967).
8. Vladimirskii, Yu. N., On strictly-cosingular operators, Dokl. Akad. Nauk SSSR, 174(6) : 1251-1252 (1967).
9. Vladimirskii, Yu. N., Φ_+-operators in locally-convex spaces, Usp. Mat. Nauk, 23(3) : 175-176 (1968).
10. Vladimirskii, Yu. N., On strictly-cosingular operators, Liet. Mat. Rinkinys, Litov. Mat. Sb., 7(3) : 399-403 (1968).
11. Gel'fond, A. O., Growth of the eigenvalues of homogeneous integral equations
12. Gol'denshtein, L. S., Gokhberg, I. Ts. and Markus, A. S., Investigation of certain properties of linear bounded operators in connection with their q-norm, Uch. Zap. Kishinev. Univ., 29 : 29-36 (1957).
13. Gol'denshtein, L. S. and Markus, A. S., On the noncompactness measure of bounded sets and linear operators, in: Research on Algebra and Mathematical Analysis, Kartya Moldovenyaske, Kishinev (1965), pp. 45-54.
14. Gol'dman, M. A., On the stability of the normal solvability properties of linear equations, Dokl. Akad. Nauk SSSR, 100(2) : 201-204 (1955).
15. Gol'dman, M. A. and Krachkovskii, S. N., On the null elements of a linear operator in its Fredholm domain, Dokl. Akad. Nauk SSSR, 86(1) : 15-17 (1952).
16. Gol'dman, M. A. and Krachkovskii, S. N., On the Riesz-Schauder operators, Usp. Mat. Nauk, 14(6) : 155-164 (1959).
17. Gol'dman, M. A. and Krachkovskii, S. N., On the Riesz-Schauder operators in locally-convex spaces, in: Functional Analysis and Its Applications, Akad. Nauk Azerbaidzhan. SSR, Baku (1961), pp. 34-35.
18. Gol'dman, M. A. and Krachkovskii, S. N., Invariance of certain spaces associated with the operator $A-\lambda I$, Dokl. Akad. Nauk SSSR, 154(3) : 500-502 (1964).
19. Gol'dman, M. A. and Krachkovskii, S. N., Perturbations of a closed linear operator, Dokl. Akad. Nauk SSSR, 158(3) : 507-509 (1964).
20. Gol'dman, M. A. and Krachkovskii, S. N., On the d-characteristic of a linear operator, Dokl. Akad. Nauk SSSR, No. 3, pp. 476-478 (1965).
21. Gol'dman, M. A. and Krachkovskii, S. N., Perturbation of homomorphisms by operators of finite rank, Dokl. Akad. Nauk SSSR, 174(4) : 743-746 (1967).
22. Gol'dman, M. A. and Krachkovskii, S. N., On the products, powers, and restrictions of homomorphisms, Dokl. Akad. Nauk SSSR, 181(5) : 1038-1041 (1968).
23. Gokhberg, I. Ts., On linear equations in normed spaces, Dokl. Akad. Nauk SSSR, 76(4) : 477-480 (1951).
24. Gokhberg, I. Ts., On linear equations in a Hilbert space, Dokl. Akad. Nauk SSSR, 76(1) : 9-12 (1951).
25. Gokhberg, I. Ts., On the zeros and null elements of unbounded operators, Dokl. Akad. Nauk SSSR, 101(1) : 9-12 (1955).
26. Gokhberg, I. Ts., On the index, null elements, and elements of the nuclei of an unbounded operator, Usp. Mat. Nauk, 12(1) : 177-179 (1957).

27. Gokhberg, I. Ts., Some properties of normally-solvable operators, Dokl. Akad. Nauk SSSR, 104(1): 9-11 (1955).
28. Gokhberg, I. Ts. and Zambitskii, M. K., Normally-solvable operators in spaces with two norms, Izv. Akad. Nauk Moldav. SSR, Ser. Fiz.-Mat. Tekh. Nauk, No. 6, pp. 80-84 (1964).
29. Gokhberg, I. Ts. and Zambitskii, M. K., On the theory of linear operators in spaces with two norms, Ukrainsk. Mat. Zh., 18(1): 11-23 (1966).
30. Gokhberg, I. Ts. and Krein, M. G., Fundamental postulates of the defect numbers, root numbers, and indices of linear operators, Usp. Mat. Nauk, 12(2): 43-118 (1957).
31. Gokhberg, I. Ts. and Krein, M. G., On completely-continuous operators with spectrum concentrated at zero, Dokl. Akad. Nauk SSSR, 128(2): 227-230 (1959).
32. Gokhberg, I. Ts. and Krein, M. G., Introduction to the Theory of Linear Nonself-Adjoint Operators, Nauka, Moscow (1965), 448 pages.
33. Gokhberg, I. Ts., Markus, A. S. and Fel'dman, I. A., On normally-solvable operators and their associated ideals, Izv. Moldav. Fil. Akad. Nauk SSSR, No. 10 (76), pp. 51-70 (1960).
34. Goursat, E. J. B., A Course in Mathematical Analysis, Ginn, Boston-New York (1917).
35. Domshlak, Yu. I., A criterion for the existence of the Fredholm alternative, Dokl. Akad. Nauk Azerbaidzhan. SSR, 14(11): 839-842 (1958).
36. Dynin, A. S., On the Fredholm theory in Banach spaces, in: Functional Analysis and Its Applications, Akad. Nauk Azerbaidzhan. SSR, Baku (1961), p. 57.
37. Keldysh, M. V., On the eigenvalues and eigenfunctions of certain classes of nonself-adjoint equations, Dokl. Akad. Nauk SSSR, 77(1): 11-14 (1951).
38. Krachkovskii, S. N., Canonical representation of the null elements of a linear operator in its Fredholm domain, Dokl. Akad. Nauk SSSR, 88(2): 201-204 (1953).
39. Krachkovskii, S. N., On the properties of a linear operator in connection with its generalized Fredholm domain, Dokl. Akad. Nauk SSSR, 91(5): 1011-1013 (1953).
40. Krachkovskii, S. N., On the expansion of the singularity domain of the operators $T_\lambda = E - \lambda A$, Dokl. Akad. Nauk SSSR, 96(6): 1101-1104 (1954).
41. Krein, M. G., Krasnosel'skii, M. A. and Mil'man, D. P., Defect numbers of linear operators in a Banach space and some geometric problems, Trudy Inst. Mat. Akad. Nauk Ukrain. SSR, 11: 97-112 (1948).
42. Krupnik, N. Ya., Stability of the index of a linear operator, Uch. Zap. Kishinev. Univ., 54: 13-19 (1960).
43. Livshits, M. S., Spectral decomposition of linear nonself-adjoint operators, Mat. Sb., 34(1): 145-198 (1954).
44. Markus, A. S., A property of the nucleus of a linear operator, Uch. Zap. Kishinev. Univ., 29: 25-28 (1957).
45. Markus, A. S., On holomorphic operator functions, Dokl. Akad. Nauk SSSR, 119(6): 1099-1102 (1958).
46. Markus, A. S. and Fel'dman, I. A., Some properties of linear operators in locally-convex spaces, Uch. Zap. Bel'ts. Gospedinst., No. 1, pp. 21-28 (1958).

47. Markus, A. S. and Fel'dman, I. A., Bounded operators in locally-convex spaces, Izv. Moldav. Fil. Akad. Nauk SSSR, 10(76): 71-78 (1960).
48. Mikhlin, S. G., On the convergence of Fredholm series, Dokl. Akad. Nauk SSSR, 42(9): 387-390 (1944).
49. Mikhlin, S. G., Multidimensional Singular Integrals and Integral Equations, Fizmatgiz, Moscow (1962), 254 pages.
50. Muskhelishvili, N. I., Singular Integral Equations, Fizmatgiz, Moscow (1962), 599 pages.
51. Nikol'skii, S. M., Linear equations in linear normed spaces, Izv. Akad. Nauk SSSR, Ser. Mat., 7(3): 147-166 (1943).
52. Panaioti, B. N., On the theory of linear singular equations in a unitary ring, Trudy Inst. Fiz. Mat., Akad. Nauk Azerbaidzhan. SSR, Ser. Mat., 3: 29-31 (1948).
53. Paraska, V. I., A metric in a set of closed operators and its application in perturbation theory, Izv. Akad. Nauk Moldav. SSR, Ser. Fiz.-Mat. Tekh. Nauk, 6: 91-96 (1964).
54. Paraska, V. I., A theorem on closed operators in a space with two norms, Izv. Akad. Nauk Moldav. SSR, Ser. Fiz.-Tekh. Mat. Nauk, No. 7, pp. 82-85 (1966).
55. Presdorf, Z., On linear equations in spaces of fundamental and generalized functions, Dokl. Akad. Nauk SSSR, 166(4): 802-805 (1966).
56. Raikov, D. A., On the complete continuity of an adjoint operator, Dokl. Akad. Nauk SSSR, 119(3): 446-449 (1958).
57. Raikov, D. A., Some properties of completely-bounded linear operators, Uch. Zap. Moskov. Gos. Ped. Inst. Lenin., 188: 171-191 (1962).
58. Samko, S. G., General singular equation on an open contour and the generalized Abel equation, Dokl. Akad. Nauk SSSR, 177(1): 44-47 (1967).
59. Sikorski, R., On determinants in a Banach space, Proc. Third All-Union Math. Congr., 1956, Vol. 4, Akad. Nauk SSSR, Moscow (1959), p. 188.
60. Fishman, K. M. and Valitskii, Yu. N., Applicability of the Fredholm theory to certain linear topological spaces, Dokl. Akad. Nauk SSSR, 117(6): 943-946 (1957).
61. Fishman, K. M. and Valitskii, Yu. N., Correction, Dokl. Akad. Nauk SSSR, 122(6): 166 (1958).
62. Khalilov, Z. I., Linear singular equations in normed rings, Dokl. Akad. Nauk SSSR, 58(8): 1613-1616 (1947).
63. Khalilov, Z. I., Linear singular equations in a normed ring, Dokl. Akad. Nauk SSSR, 60(7): 1133-1136 (1948).
64. Khalilov, Z. I., Linear singular equations in a normed ring, Izv. Akad. Nauk SSSR, Ser. Mat., 13(2): 163-176 (1949).
65. Cherskii, Yu. I., The general singular equation and equation of the convolution type, Mat. Sb., 41(3): 277-296 (1957).
66. Shvarts, A. S., On the homotopic topology of Banach spaces, Dokl. Akad. Nauk SSSR, 154(1): 61-63 (1964).
67. Shubin, M. A., Factorization of parameter-dependent matrix functions in normed rings and related aspects of the theory of Noether operators, Mat. Sb., 73(4): 610-629 (1967).

68. Altman, M., On linear functional equations in locally-convex linear topological spaces, Studia Math., 13(2): 194-207 (1953).
69. Altman, M., The Fredholm theory of linear equations in locally-convex linear topological spaces, Bull. Acad. Polon. Sci., 2(6): 267-269 (1954).
70. Altman, M., On linear functional equations in (B_0)-spaces, Studia Math., 15(2): 131-135 (1956).
71. Atkinson, F., A spectral problem for completely-continuous operators, Acta Math. Acad. Sci. Hung., 3: 53-60 (1952).
72. Atkinson, F., On relatively-regular operators, Acta Sci. Math., 15(1): 38-56 (1953).
73. Audin, M., On the linear transformations of vector spaces satisfying a condition of Fredholm [in French], Compt. Rend., 244(6): 711-713 (1957).
74. Audin, M., Linear transformations that verify a condition of Fredholm, and the spectrum of certain families of transformations [in French], Compt. Rend., 244(24): 2880-2882 (1957).
75. Avantaggiati, A., Theorems on linear transformations in Banach spaces [in Italian], Ric. Mat., 16(1): 116-126 (1967).
76. Banach, S., On functionals [in French], Studia Math., 1: 223-239 (1929).
77. Barnes, B. A., A generalized Fredholm theory for certain maps in the regular representations of an algebra, Canad. J. Math., 20(2): 495-504 (1968).
78. Breuer, M., Banach Algebras with Applications to Fredholm Operators and Singular Integral Equations [in German], Bonner Math., Schr., No. 24, Bonn (1965), 108 pages.
79. Breuer, M. and Cordes, H. O., On Banach algebras with σ-symbol (I), J. Math. Mech., 13: 313-323 (1964).
80. Breuer, M. and Cordes, H. O., On Banach algebras with σ-symbol (II), J. Math. Mech., 14: 299-314 (1965).
81. Browder, F. E., Functional analysis and partial differential equations (I), Math. Ann., 138(1): 55-79 (1959); Russian translation in: Matematika, Sb. Perev. Inostr. Statei, 4(3): 79-106 (1960).
82. Buraczewski, A., The determinant theory of generalized Fredholm operators, Studia Math., 22(3): 79-106 (1960).
83. Buraczewski, A., On a certain property of determinant systems, Colloq. Math., 10(2): 325-330 (1963).
84. Caradus, S. R., Operators of Riesz type, Pacif. J. Math., 18(1): 61-71 (1966).
85. Caradus, S. R., Operators with finite ascent and descent, Pacif. J. Math., 18(3): 437-449 (1966).
86. Caradus, S. R., On meromorphic operators (I), Canad. J. Math., 19(4): 723-736 (1967).
87. Caradus, S. R., On meromorphic operators (II), Canad. J. Math., 19(4): 737-748 (1967).
88. Caradus, S. R., On Fredholm operators (abstract), Queen's Papers Pure Appl. Math., No. 10, pp. 220-225 (1967).
89. Carleman, T., On the theory of linear integral equations [in German], Math. Z., 9: 196-217 (1921).

90. Coburn, L. A. and Lebow, A., Approximation by Fredholm operators in the metric space of closed operators, Rend. Semin. Mat. Univ. Padova, 36(2): 217-222 (1966).
91. Coburn, L. A. and Lebow, A., Algebraic theory of Fredholm operators, J. Math. Mech., 15(4): 557-584 (1966).
92. Cordes, H. O., A nonalgebraic characterization of the $\mathfrak{T}$-Fredholm operators [in German], Math., Ann., 163(3): 212-229 (1966).
93. Cordes, H. O., On a generalized Fredholm theory, J. Reine Angew. Math., 227: 121-149 (1967).
94. Cordes, H. O. and Labrousse, J. P., The invariance of the index in the metric space of closed operators, J. Math., Mech., 12(5): 693-719 (1963).
95. Deprit, A., Riesz endmorphisms [in French], Ann. Soc. Sci. Bruxelles, Ser. 1, 70(3): 165-183 (1956).
96. Deprit, A., Some classes of homomorphisms in vector spaces [in French], Ann. Soc. Sci. Bruxelles, Ser. 1, 71(1): 5-43 (1957).
97. Deprit, A., Some classes of homomorphisms of separable locally-convex spaces [in French], Bull. Cl. Sci. Acad. Roy. Belg., 43(4): 252-272 (1957).
98. Deprit, A., Some classes of endomorphisms of separable locally-convex spaces [in French], Ann. Soc. Sci. Bruxelles, Ser. 1, 71(2): 89-101 (1957).
99. Deprit, A., A type of homomorphisms of separable locally-convex spaces [in French], Ann. Soc. Sci. Bruxelles, Ser. 1, 72(1): 5-13 (1958).
100. Deprit, A., Contribution to the investigation of the algebra of continuous linear operators of a separable locally-convex space [in French], Mem. Acad. Roy. Belg., Cl. Sci., Coll. Vol. 8, 31, fasc. 2, 170S.
101. Dieudonne, J. and Schwartz, L., Duality in the spaces (F) and (LF) [in French], Ann. Inst. Fourier, 1: 61-101 (1949).
102. Douady, A., A Banach space whose linear group is not connected [in French], Proc. Koninkl. Nederland. Akad. Wet., A68(5): 787-789 (1965); Indagationes Math., 27(5): 787-789 (1965).
103. Föster, K.-H., On the invariance of spaces associated with the operator $T - \lambda A$ [in German], Arch. Math., 17(1): 56-64 (1966).
104. Förster, K.-H., On linear closed operators analytically dependent on a parameter [in German], Math., Z., 95(4): 251-258 (1967).
105. Fredholm, I., On a class of functional equations [in French], Acta Math., 134: 1561-1564 (1902).
106. Fredholm, I., On a class of functional equations [in French], Acta Math., 27: 365-390 (1903).
107. Gamelin, T., Decomposition theorems for Fredholm operators, Pacif. J. Math., 15(1): 97-106 (1965).
108. Gindler, H. A. and Taylor, A. E., The minimum modulus of a linear operator and its use in spectral theory, Studia Math., 22(1): 15-41 (1962).
109. Goldberg, S., Unbounded Linear Operators; Theory and Applications, New York (1966).
110. Goldberg, S. and Thorp, E., On some open questions concerning strictly-singular operators, Proc. Amer. Math. Soc., 14(2): 334-336 (1963).

111. Gramsch, B., Integration and holomorphic functions in locally-bounded spaces [in German], Math. Ann., 162:190-210 (1965).
112. Gramsch, B., σ-Transformations in locally-bounded vector spaces [in German], Math., Ann., 165(2):135-151 (1966).
113. Gramsch, B., A scheme for the theory of Fredholm endomorphisms and an application to the ideal chains of Hilbert spaces [in German], Math. Ann., 171(4): 263-272 (1967).
114. Gramsch, B., An ideal structure of Banach operator algebras [in German], J. Reine Angew. Math., 225:97-115 (1967).
115. Graves, C. M., A generalization of the Riesz theory of completely-continuous transformations, Trans. Amer. Math. Soc., 79(1):141-149 (1955).
116. Grothendieck, A., On the notion of the topological tensor product of topological vector spaces, and a remarkable class of vector spaces associated with that notion [in French], Compt. Rend., 233:1556-1558 (1951).
117. Grothendieck, A., Topological tensor products and nuclear spaces [in French], Mem. Amer. Math. Soc., No. 16, p. 140 (1955).
118. Grothendieck, A., Correction to the paper, "Topological tensor products and nuclear spaces" [in French], Ann. Inst. Fourier, 6:117-120 (1955-56).
119. Grothendieck, A., The Fredholm theory [in French], Bull. Soc. Math., France, 84(4):319-384 (1956).
120. Gruson, L., The p-Fredholm theory [in French], Bull. Soc. Math. France, 94(1):67-95 (1966).
121. Hellinger, E. and Toeplitz, O., Integral equations and equations with infinitely-many unknowns [in German], Encyklopädie Math. Wiss, Vol. II, C13, pp. 1335-1616 (1928).
122. Heuser, H., On the eigenvalue theory of a class of Riesz operators [in German], Arch. Math., 14(1):39-46 (1963).
123. Fukuhara, M. and Sibuya, Y., On the completely-continuous endomorphisms [in French], Proc. Japan. Acad., 31(9):595-599 (1955).
124. Fukuhara, M. and Sibuya, Y., Theory of completely-continuous endomorphisms (I) [in French], J. Fac. Sci. Univ. Tokyo, Sec. I, 7(4):391-405 (1957).
125. Fukuhara, M. and Sibuya, Y., Theory of completely-continuous endomorphisms (II) [in French], J. Fac. Sci. Univ. Tokyo, Sec. I, 7(5):511-525 (1958),
126. Kaashoek, M. A., Closed linear operators on Banach spaces, Proc. Koninkl. Nederland. Akad. Wet., A68(3):405-414 (1965); Indagationes Math., 27(3): 405-414 (1965).
127. Kaashoek, M. A., Stability theorem for closed linear operators, Proc. Koninkl. Nederland. Akad. Wet., A68(3):452-466 (1965); Indagationes Math., 27(3): 452-466 (1965).
128. Kaashoek, M. A., Ascent, descent, nullity, and defect, a note on a paper by A. E. Taylor, Math. Ann., 172(2):105-115 (1967).
129. Kaashoek, M. A., On the Riesz set of a linear operator, Proc. Koninkl. Nederland. Akad. Wet., A71(1):46-53 (1968); Indagationes Math., 30(1); 46-53 (1968).
130. Kaashoek, M. A. and Lay, D. C., On operators whose Fredholm set is the complex plane, Pacif. J. Math., 21(2):275-278 (1967).

131. Kaniel, S. and Schechter, M., Spectral theory for Fredholm operators, Commun. Pure Appl. Math., 16(4): 423-448 (1963).

132. Kato, T., Perturbation theory for nullity, deficiency, and other quantities of linear operators, J. Analise Math., 6(2): 261-322 (1958).

133. Kato, T., Perturbation Theory for Linear Operators, Springer, Berlin (1966), 592 pages.

134. Kleinecke, D., Almost-finite, compact, and inessential operators, Proc. Amer. Math. Soc., 14(6): 863-868 (1963).

135. Köthe, G., On the theory of compact operators in locally-convex spaces [in German], Portugal. Math., 13(1): 97-104 (1954).

136. Köthe, G., General linear transformations of locally-convex spaces, Math. Ann., 159(5): 309-328 (1965).

137. Köthe, G., Linear Topological Spaces [in German], Vol. I, Springer, Berlin (196C

138. Kuiper, N., The homotopy type of the unitary group of Hilbert space, Topology, 3(1): 19-30 (1965).

139. Kultze, R., On the theory of Fredholm endomorphisms in nuclear topological vector spaces [in German], J. Reine Angew. Math., 200(1-2): 112-124 (1958).

140. Lacey, H., Generalized compact operators in locally-convex spaces, Doct. Dissert., New Mexico State Univ., (1963), 53 pages; Dissert. Abstr., 24(3): 1187 (1963).

141. Lacey, H. and Whitley, K., Conditions under which all the bounded linear maps are compact., Math. Ann., 158(1): 1-5 (1965).

142. Leray, J., Eigenvalues and eigenvectors of a completely-continuous endomorphism of a vector space in convex neighborhoods [in French], Acta Sci. Math. Szeged., 12: 177-186 (1950).

143. Leżański, T., The Fredholm theory of linear equations in Banach spaces, Studia Math., 13(2): 244-276 (1953).

144. Leżański, T., On multiplicative functionals [in French], Studia Math., 14(1): 13-23 (1953).

145. Luft, E., Maximal R-sets, Grassmann spaces, and Stiefel spaces of a Hilbert space, Trans. Amer. Math. Soc., 126(1): 73-107 (1967).

146. Nagy, B., (Szökefalvi-Nagy), On the stability of the index of unbounded linear transformations, Acta Math. Acad. Hung., 4: 49-51 (1952).

147. Nakamura, M., On operators of Schaefer class in the theory of singular integral equations, Proc. Japan. Acad. 33(8): 455-456 (1957).

148. Neubauer, G., On the index of closed operators in Banach spaces (I) [in German], Math. Ann., 160(2): 93-130 (1965).

149. Neubauer, G., On the index of closed operators in Banach spaces (II) [in German], Math., Ann., 162(1): 92-119 (1965).

150. Nieto, J., On Fredholm operators and the essential spectrum of singular integral operators, Math., Ann., 178(1): 62-77 (1968).

151. Noether, E., On a class of singular integral equations [in German], Math. Ann., 82: 42-63 (1921).

152. Clagunju, P. A. and West, T. T., The spectra of Fredholm operators in locally-convex spaces, Proc. Cambridge Phil. Soc., 60(4): 801-807 (1964).

153. Oliver, R. K., Note on a duality relation of Kaashoek, Proc. Koninkl. Nederland. Akad. Wet., A69(3): 364-368 (1966).

154. Pelczyński, A., On strictly-singular and strictly-cosingular operators, I: Strictly-singular and strictly-cosingular operators in C (S)-spaces, Bull. Acad. Polon. Sci., Ser. Math. Astron. Phys., 13(1): 31-36 (1965).
155. Pelczyński, On strictly-singular and strictly-cosingular operators, II: Strictly-singular and strictly-cosingular operators in L (ν)-spaces, Bull. Acad. Polon. Sci., Ser. Math. Astron. Phys., 13(1): 37-41 (1965).
156. Pettineo, B., On the alternative theorem for certain functional equations [in Italian], Atti Accad. Naz. Lincei. Rend. Cl. Sci. Fis. Mat. Natur., 30(5): 694-699 (1961).
157. Pettineo, B., On the minimal dimension of the proper spaces of second-order equations in Hilbert spaces [in Italian], Atti Accad. Naz. Lincei Rend. Cl. Sci. Fis. Mat. Natur, 38(2): 150-154 (1965).
158. Pietsch, A., On the theory of σ-transformations in locally-convex vector spaces [in German], Math. Nachr., 21(6): 347-369 (1960).
159. Pietsch, A., A generalized spectral problem for compact linear representations in locally-convex vector spaces [in German], Math., Ann., 140(2): 147-152 (1960).
160. Pietsch, A., Unstable linear representations in locally-convex vector spaces [in German], Math. Ann., 140(2): 153-164 (1960).
161. Pietsch, A., Homomorphisms in locally-convex vector spaces [in German], Math. Nachr., 22(3-4): 162-174 (1960).
162. Pietsch, A., Quasi-precompact endomorphisms and an ergodic principle in locally-convex vector spaces [in German], J. Reine Angew. Math., 207-(1-2): 16-30 (1961).
163. Pietsch, A., On the Fredholm theory in locally-convex spaces [in German], Studia Math., 22(2): 161-179 (1963).
164. Przeworska-Rolewicz, D., On involutions of order n [in French], Bull. Acad. Polon. Sci., Ser. Sci. Math. Astron. Phys., 8(11-12): 735-739 (1960).
165. Przeworska-Rolewicz, D., On involutive equations of order n [in French], Bull. Acad. Polon. Sci., Ser. Sci. Math. Astron. Phys., 8(11): 741-746 (1960).
166. Przeworska-Rolewicz, D., Involutive equations and their operators [in French], Studia Math., 20(2): 95-117 (1961).
167. Przeworska-Rolewicz, D., Operators satisfying a polynomial identity [in French], Studia Math., 22(1): 43-58 (1962).
168. Przeworska-Rolewicz, D., Equations with algebraic operators [in French], Studia Math., 22(3): 337-367 (1963).
169. Przeworska-Rolewicz, D., Equations with almost-algebraic operators [in French], Studia Math., 25(2): 163-180 (1965).
170. Przeworska-Rolewicz, D. and Rolewicz, S., Remarks on Φ-operators in linear topological spaces, Roczn. Pols. Towarz. Mat., Ser. 1, 9(1): 91-94 (1965).
171. Przeworska-Rolewicz, D. and Rolewicz, S., On operators preserving a conjugate space, Studia Math., 25(2): 245-249 (1964/65).
172. Przeworska-Rolewicz, D. and Rolewicz, S., On quasi-Fredholm ideals, Studia Math., 26(1): 67-71 (1965).
173. Przeworska-Rolewicz, D. and Rolewicz, S., On d-characteristic and d_{Ξ}-characteristic of linear operators, Ann. Polon. Math., 19(2): 117-121 (1967).

174. Riexz, F., On linear functional equations [in German], Acta Math., 41 : 71-98 (1916).
175. Ringrose, J. R., Precompact linear operators in locally-convex spaces, Proc. Cambridge Phil. Soc., 53(3): 581-591 (1957).
176. Ruston, A. F., Direct product of Banach spaces and linear functional equations, Proc. London Math. Soc., Ser. 3, 1 : 327-387 (1951).
177. Ruston, A. F., On the Fredholm theory of integral equations for operators belonging to trace class of a general Banach space, Proc. London Math. Soc., 1 : 109-124 (1951).
178. Ruston, A. F., Operators with a Fredholm theory, J. London Math. Soc., 29(3): 318-326 (1954).
179. Ruston, A. F., Formulae of Fredholm type for compact linear operators on a general Banach space, Proc. London Math. Soc., 3(11): 368-377 (1953).
180. Ruston, A. F., Fredholm formulae and the Riesz theory, Compositio Math., 18(1-2): 25-48 (1966).
181. Saphar, P., Nuclear power applications and Hilbert-Schmidt applications [in French], Ann. Sci. Ecole Norm. Super., 83(2 : 113-151 (1966/67).
182. Schaefer, H. H., On singular integral equations and a class of homomorphisms in locally-convex spaces [in German], Math., Z., 66(2): 147-163 (1956).
183. Schaefer, H. H., On the Fredholm alternative in locally-convex linear spaces, Studia Math., 18(3): 229-245 (1959).
184. Schatten, R., A Theory of Cross-Spaces, Princeton Univ. Press (1950).
185. Schauder, G., On linear completely-continuous functional operators [in German], Studia Math., 2 : 183-196 (1930).
186. Schechter, M., Basic theory of Fredholm operators, An. Scuola Norm. Super. Pisa Sci. Fis. Mat., 21(2): 261-280 (1967).
187. Schwartz, L., Homomorphisms and completely-continuous applications [in French], Compt. Rend., 236(26): 2472-2473 (1953).
188. Shirota, T., On completely-continuous operators on locally-convex vector spaces, Proc. Japan. Acad., 30(9): 837-840 (1954).
189. Sikorski, R., On Leżański's determinants of linear equations in Banach spaces, Studia Math., 14(1): 24-48 (1953).
190. Sikorski, R., On determinants of Leżański and Ruston, Studia Math., 16(2): 99-112 (1957).
191. Sikorski, R., Determinant systems, Studia Math., 18(2): 161-186 (1959).
192. Sikorski, R., On Leżański endomorphisms, Studia Math., 18(2): 187-189 (1959).
193. Sikorski, R., The determinant theory of the Carleman type, Bull. Acad. Polon. Sci., Ser. Sci. Math. Astron. Phys., 8(10): 685-689 (1960).
194. Sikorski, R., Remarks on Leżański's determinants, Studia Math., 20(2): 145-161 (1961).
195. Sikorski, R., The determinant theory in Banach spaces, Colloq. Math., 8(2): 141-198 (1961).
196. Sikorski, R., On the Carleman determinants, Studia Math., 20(3): 327-346 (1961).
197. Sikorski, R., Determinants in Banach spaces, Studia Math., No. 1, Ser. Spec., pp. 111-116 (1963).

198. Smithies, F., The Fredholm theory of integral equations, Duke Math. J., 8:107-130 (1941).
Szökefalvi-Nagy, B., (*see* Nagy, B. [146]).
199. Taylor, A., Theorems on ascent, descent, nullity, and defect of linear operators, Math. Ann., 163(1):18-49 (1966).
200. Togo, S. and Shiraishi, R., Note on F-operators in locally-convex spaces, J. Sci. Hiroshima Univ., Ser. A., Div. I, 29(2):243-251 (1961).
201. Wendland, W., The Fredholm alternative for operators which are adjoint with respect to a bilinear functional [in German], Math. Z., 101(1):61-64 (1967).
202. West, T. T., The decomposition of Riesz operators, Proc. London Math., Soc., 16(4):737-752 (1966).
203. West, T. T., Riesz operators in Banach spaces, Proc. London Math. Soc., 16(1): 131-140 (1966).
204. Whitley, R. J., Strictly-singular operators and their conjugates, Trans. Amer. Math.Soc., 113(2):252-261 (1964).
205. Williamson, J. H., Compact linear operators in linear topological spaces, J. London Math. Soc., 29 (Part 2):149-156 (1954).
206. Yood, B., Difference algebras of linear transformations on a Banach space, Pacif. J. Math., 4(4):615-636 (1954).
207. Roshida, K., Quasi-completely-continuous linear functional equations, Japan. J. Math., 15:247-301 (1939).

Supplement to Literature Cited

1*. Vladimirskii, Yu. N., Φ_--operators in locally-convex spaces, Dokl. Akad. Nauk SSSR, 184(3):514-517 (1969).
2*. Dikanskii, A. S., Certain classes of mappings in locally-convex spaces, Izv. Akad. Nauk Azerbaidzhan. SSR, Fiz.-Tekh. Mat. Nauk, No. 3 (1969).
3*. Breuer, M., Fredholm theories on von Neumann algebras (I), Math. Ann., 178(3): 234-254 (1968).
4*. de Bruyn, G. F. C., Asymptotic properties of linear operators, Proc. London Math. Soc., 18(3):405-427 (1968).
5*. Gillespie, T. A. and West, T. T., A characterisation and two examples of Riesz operators, Glasgow Math. J., 9(2):106-110 (1968).
6*. Köthe, G., The image spaces of closed operators [in German], J. Reine Angew. Math., 232:110-111 (1968).
7*. Neubauer, G., Homotopy properties of semi-Fredholm operators in Banach spaces, Math. Ann., 176(4):273-301 (1968).
8*. Webb, J. H., Perturbation theory for a linear operator, Proc. Cambridge Phil. Soc., 631:11-20 (1967).
9*. Webb, J. H., Linear operators with closed range, Proc. Cambridge Phil. Soc., 64(4):1009-1010 (1968).

Representations of Groups and Algebras in Spaces with Indefinite Metric

M. A. Naimark and R. S. Ismagilov

Introduction

Let H be a Hilbert space with scalar product [x, y], and let P_+, P_- be orthogonal projectors in H, where $P_+ + P_- = E$; we set $J + P_+ - P_-$, $H_+ = P_+ H$, and $H_- = P_- H$. In H we defined the bilinear form

$$(x, y) = [Jx, y], \tag{1}$$

which we call the indefinite scalar product. If $k = \min(\dim H_+, \dim H_-)$, then H [assigned form (1)] is said to be a space of type Π_k, a well as a J-space; we also say that form (1) determines in H the Π_k-metric, or indefinite metric. We shall have frequent occasion to denote the space H by Π_k. If $k < \infty$, then Π_k is called a Pontryagin space (after L. S. Pontryagin). In this case we shall assume that $k = \dim H_+$ (this can always be accomplished, of course, by commutation of the operators P_+ and P_-).

An element $x \in H$ is said to be nonnegative if $(x, x) \geq 0$ and positive if $(x, x) > 0$. A set $S \subset H$ is said to be nonnegative if S consists only of nonnegative elements and positive if any nonvanishing element $x \in S$ is positive. The notions of nonpositive and negative elements (or sets) are defined analogously. Finally, an element $x \in H$ is said to be neutral if $(x, x) = 0$; a set consisting only of neutral elements is said to be neutral.

Henceforth we consider all topological concepts in the sense of the Hilbert norm $\|x\| = \sqrt{[x, x]}$. On the other hand, the concepts of the orthogonality of vectors and subspaces of H and of isometric, unitary, and self-adjoint operators are defined with respect to the form (x, y) as with respect to the conventional scalar product. Ordinarily in this case the terms "J-orthogonal vectors," "J-unitary operator," etc., are used; however, we shall omit the letter J.

Problems bearing on the geometry of J-spaces and the theory of unitary and Hermitian operators in J-spaces have been investigated in several papers; we shall list some of them below. Our principal concern in the present article is with the theory of representations of groups and algebras in J-spaces.

Let G be a topological group, and H a space of type Π_k $(0 < k \leq \infty)$. A representation of G in H unitary in the Π_k-metric is defined as the mapping $g \to U_g$ of G into a group of bounded and invertible operators in H satisfying the following conditions:

1) U_g is unitary in the Π_k-metric for every $g \in G$;

2) $U_e = E$, where e is the unit element of the group and E is the unit operator in H;

3) $U_{g_1 g_2} = U_{g_1} U_{g_2}$ for $g_1, g_2 \in G$;

4) $(U_g x, y)$ is a continuous numerical function on G for any $x, y \in H$..

Analogously, a symmetric representation in a space H of type Π_k of a normed algebra A with involution $a \leftrightarrow a^*$ is defined as a mapping $a \to U_a$ invested with the following properties:

1) U_a is a bounded linear operator in H for any $a \in A$;

2) $\| U_a \| \leq c \|a\|$, where $\| U_e \|$ is the norm of the operator U_a in H and $\|a\|$ is the norm of the element a in A;

3) $U_{\alpha_1 a_1 + \alpha_2 a_2} = \alpha_1 U_{a_1} + \alpha_2 U_{a_1}$, $U_{a_1 a_2} = U_{a_1} U_{a_2}$ for any $a_1, a_2 \in A$ and any complex numbers α_1, α_2;

4) $U_{a^*} = (U_a)^*$ for every $a \in A$, where $(U_a)^*$ denotes the adjoint of the operator U_a under the Π_k-metric.

A symmetric representation in a space of type Π_k is also said to be symmetric in the Π_k-metric.

Fairly comprehensive results in the description of the representations have been gained for the case $0 < k < \infty$ (i.e., for Pontryagin spaces) (apropos, see below, § 1, Sec. 1.1). As in the case of representations in a Hilbert space, the general investigation of the representations defined above rests on the preliminary study of commutative symmetric algebras in spaces of type Π_k. This is the subject of § 2. In preparation for the latter, in § 1 we consider some theorems on invariant subspaces, which are important in the ensuing sections.

The representations of groups are discussed in § 3. In § 4 we present the results of Phillips on the extensions of subspaces invariant under an operator algebra, as well as the applications of those results to differential equations and dissipative operators.

§ 1. Theorems on Invariant Subspaces

1.1. A nonnegative subspace $\mathfrak{M}$ of a space H with Π_k-metric is called a maximal nonegative subspace if in H there does not exist a nonnegative subspace containing $\mathfrak{M}$ as a regular part thereof. The maximal nonpositive subspace is defined analogously.

If $0 < k < \infty$, a subspace $\mathfrak{M}$ is maximal nonnegative if and only if $\mathfrak{M} = k$. Spaces of type Π_k with $0 < k < \infty$ were first studied by Pontryagin [37] under the influence of Sobolev's work [38] on applied problems.* Pontryagin's fundamental result is contained in the following:

Theorem 1. For any self-adjoint operator A in a space of type Π_k $(0 < k < \infty)$ there exists a maximal (and, hence, k-dimensional) nonnegative subspace invariant under A.

The analogous statement holds for a unitary operator [5]. M. G. Krein has shown that for k = 1 these results can be deduced by the application of topological fixed-point theorems [13]. In [5] Krein and Iokhvidov extended the same method to the case of a space of type Π_k with any $k < \infty$. In [5, 6] the reader will also find a detailed study of the theory of type Π_k, including analogs of Theorem 1 and its applications to the spectral theory of operators in spaces with Π_k-metric; the cited papers also contain extensive bibliographies.

In the case of spaces of type Π_∞ an analogous result has been obtained by Langer [46] under an additional limitation on the self-adjoint operator A. For unitary operators a similar result has been obtained by Krein [14].

*Sobolev carried out this study in 1944 and published his results in 1960.

Later Fan [43] proved a topological theorem from which is inferred the analog of Theorem 1 for a unitary operator; in a footnote to the Russian edition of [43] I. S. Iokhvidov shows that this theorem can be used to deduce the indicated result of Krein. It is not yet known whether the assertion analogous to Theorem 1 is true in the general case (for Π_∞).

1.2. The following theorem obtained by Naimark [49] generalizes Theorem 1 (or, more precisely, its analog for unitary operators) to a family of operators.

Theorem 2. For every family $\mathfrak{A}$ of pairwise-commutating unitary operators in a space of type Π_k, $0 < k < \infty$, there exists a maximal nonnegative (and, hence, k-dimensional) subspace invariant under all operators of $\mathfrak{A}$.

From the foregoing we obtain:

Corollary 1. The statement of Theorem 2 remains valid for:

1) a family $\mathfrak{L}$ of pairwise-commutating bounded Hermitian operators in Π_k;

2) a family $\mathfrak{L}$ of pairwise-commutating bounded linear operators in Π satisfying the following condition: $A \in \mathfrak{L}$ implies $A^* \in \mathfrak{L}$. Thus, it suffices to put $\mathfrak{A} = \{e^{itA}, A \in \mathfrak{L}, -\infty < t < \infty\}$ in case 1) and $\mathfrak{A} = \{\frac{1}{2}(B + B^*), B \in \mathfrak{L}\}$ in case 2).

Theorem 2 plays the most important role in the ensuing sections. It is by virtue of this theorem that sufficiently complete results have been obtained for the case $0 < k < \infty$ with regard to the representations of groups and algebras in spaces of type Π_k. The greater part of the present article is devoted to the case $0 < k < \infty$.

Fan [44] has proved a theorem on invariant subspaces for a different case from the one considered in Theorem 2. Langer [16] has proved that, given additional assumptions concerning the operators A (which hold for $k < \infty$) the statements of Corollary 1 are also true for $k = \infty$. The analogs of Theorem 2 and its corollaries have not been proved in the general case.

Another generalization has been obtained by Ismagilov in [7], in which he analyzes the symmetric representation in a space of type Π_k ($k < \infty$) of an algebra with involution endowed with a complete family of representations of dimension not greater than m ($m < \infty$). He shows that there exists in this case an invariant subspace $H_1 \subset H$ such that $\dim H_1 \leq k \cdot m$ and the orthogonal complement $H_1^{\perp}$ of H_1 is nonpositive.

§ 2. Description of Commutative Symmetric Algebras of Operators in Spaces of Type Π_k

Let R be an algebra of bounded linear operators in a space of type Π_k. We say that R is symmetric if $A \in R$ implies $A^* \in R$. Algebras R_1 and R_2 in spaces H_1 and H_2 (each of which is of type Π) are said to be equivalent if there exists an isometric mapping of H_1 onto H_2 taking R_1 into R_2. The problem of describing all (up to equivalence) commutative symmetric algebras in spaces of type Π_k ($k < \infty$) has been solved by Naimark [30, 31, 35, 48]. This result is presented below (Secs. 2.1 and 2.2). In Sec. 2.3 we give the results of Langer [45] and Loginov [18, 19] on the analysis (and simplification) of Naimark's model.

2.1 Commutative Symmetric Algebras in Spaces of Type Π_1. Let R be a commutative symmetric algebra in a space H with Π_1-metric. By Corollary 1 there exists in H a one-dimensional nonnegative subspace $\mathfrak{N}$ invariant under R. This means that

$$Ax = \lambda x, \quad A \in R, \quad x \in \mathfrak{N},$$

where, clearly, the function $\lambda(A)$ is a homomorphism, continuous on the operator norm, of R in the field of complex numbers. We refer to $\lambda(A)$ as an eigenfunctional of R.

Only the following cases are possible:

I. Among the nonnegative one-dimensional subspace $\mathfrak{N}$ invariant under R there exists a positive subspace.

II. All nonnegative one-dimensional invariant subspaces are neutral, and among them there is an $\mathfrak{N}$ such that $\operatorname{Im}\lambda(A_0) \neq 0$ for the corresponding homomorphism $\lambda(A)$ and for some $A_0 = A_0^* \in R$.

III. All nonnegative one-dimensional invariant subspaces are neutral, and for each of them the corresponding function $\lambda(A)$ is real on every $A \doteq A^* \in R$.

We investigate each case separately.

Case I. It has been proved that $\lambda(A)$ is independent of the choice of subspace $\mathfrak{N}$ and that $\lambda(A)$ is real if $A = A^* \in R$. Let us set $\mathfrak{M} = \{x : x \in H, Ax = \lambda(A)x \text{ for all } A \in R\}$, $\mathfrak{H} = \mathfrak{M}^{\perp}$. Then $H = \mathfrak{M} \oplus \mathfrak{H}$, and $\mathfrak{M}$ and $\mathfrak{H}$ are invariant under all $A \in R$; the restriction of $A \in R$ to the subspace $\mathfrak{H}$ is denoted by A_1. Then the set $R_1 = \{A_1 : A \in R\}$ is a commutative symmetric algebra in $\mathfrak{H}$ satisfying the following condition:†

$$\textit{If } A_1 x_1 = \lambda(A) x_1 \textit{ for all } A \in R \textit{ and some } x_1 \in \mathfrak{H}, \textit{ then } x_1 = 0. \tag{2.1}$$

The following subcases are possible:

Ia. $A_1 = 0$ implies $\lambda = 0$.

Ib. There exists an operator $A^0 \in R$ such that $A_1^0 = 0$, but $\lambda(A^0) \neq 0$.

In case Ia the homomorphism $A_1 \to \lambda(A_1) = \lambda(A)$ on R_1 is uniquely defined, and condition (2.1) is written as follows:

$$A_1 x = \lambda(A_1) x \textit{ for all } A \in R \textit{ implies } x = 0. \tag{2.2}$$

In case Ib, setting $A = A^0$, we see that condition (2.2) is fulfilled automatically. A straightforward argument leads to the following theorem.

Theorem 3. In case I the algebra is defined by:

1) the space $\mathfrak{M}$, which is either one-dimensional and positive or is of type Π_1;

2) the commutative symmetric algebra R_1 in some Hilbert space $\mathfrak{H}$ and in case Ia, in addition, by the symmetric homomorphism $A_1 \to \lambda(A_1)$ or R_1 in the field of complex numbers, satisfying condition (2.2). Here R is realized as follows: H consists of all (formal) sums of the type $x = x_0 + x_1$, where $x_0 \in \mathfrak{M}$, $x_1 \in \mathfrak{H}$ with componentwise addition and multiplication by a number and its scalar product $(x, x') = (x_0 = X_0') - [x_1, x_1']$ for $x = x_0 + x_1$, $x' = x_0' + x_1'$, where (x_0, x_0') and $[x_1, x_1']$ are the scalar products in $\mathfrak{M}$ and $\mathfrak{H}$, re-

†This condition is omitted in [34]; the omission was kindly brought to the author's attention by Yu. L. Shmul'yan.

spectively. The algebra R consists of all operators A in H defined by the formula $Ax = \lambda x_0 + A_1 x_1$, $A_1 \in R_1$, where λ is an arbitrary number independent of A_1 in case Ib and $\lambda = \lambda(A_1)$ in case Ia. Two algebras R and $\tilde{R}$ corresponding to the "triples" $\mathfrak{M}$, $\mathfrak{H}$, R_1 and $\tilde{\mathfrak{M}}$, $\tilde{\mathfrak{H}}$, $\tilde{R}_1$ [and $\lambda(A_1)$, $\tilde{\lambda}(A_1)$ in case Ib] are isomorphic if and only if: 1) $\dim \mathfrak{M} = \dim \tilde{\mathfrak{M}}$; 2) R_1 and $\tilde{R}_1$ are spatially isomorphic; 3) in case Ia there exists an isometric mapping of $\mathfrak{H}$ onto $\tilde{\mathfrak{H}}$ that maps R_1 onto $\tilde{R}_1$ and takes $\lambda(A_1)$ into $\tilde{\lambda}(A_1)$.

Case II. It turns out in this case that there are exactly two one-dimensional invariant nonnegative subspaces $\mathfrak{N}$, $\mathfrak{N}'$; both are neutral and piecewise-connected,† and the corresponding homomorphisms $\lambda(A)$ and $\mu(A)$ are connected by the relation

$$\lambda(A) = \overline{\mu(A^*)}.$$

Let us set

$$\mathfrak{M} = \mathfrak{N} \dotplus \mathfrak{N}', \quad \mathfrak{H} = \mathfrak{M}^{\perp}, \quad H = \mathfrak{M} \oplus \mathfrak{H},$$

where $\mathfrak{M}$ and $\mathfrak{H}$ are invariant under R, $\mathfrak{M}$ is a two-dimensional subspace of type Π_1, and $\mathfrak{H}$ is a Hilbert space with scalar product $[x_1, x_1'] = -(x_1, x_1')$; $\mathfrak{M}$ is called the hyperbolic space of the algebra R. The restriction of $A \in R$ to $\mathfrak{H}$ is denoted by A_1.

Only the following two subcases are possible:

IIa. $A_1 = 0$ implies $\lambda(A) = 0$.

IIb. There exists an operator $A^0 \in R$ such that $A_1^0 = 0$, but $\lambda(A^0) \neq 0$.

In case IIa the functions $\lambda_1(A_1) = \lambda(A)$, $\mu(A_1) = \mu(A)$ are uniquely determined. Further analysis leads to the following theorem.

Theorem 4. In case IIb the algebra R is defined by the commutative symmetric algebra R_1 of operators in the Hilbert space $\mathfrak{H}$, and in case IIa it is defined by the algebra R_1 and homo-

†Two neutral subspaces $\mathfrak{N}$, $\mathfrak{N}'$ are said to be piecewise-connected if $\mathfrak{N}$ does not contain nonvanishing vectors orthogonal to $\mathfrak{N}'$ and if $\mathfrak{N}'$ does not contain nonvanishing vectors orthogonal to $\mathfrak{N}$.

morphism $\lambda(A)$ of R_1 in the field of complex numbers [where $\lambda(A^*) \neq \overline{\lambda(A)}$].

The algebra R is realized as follows: H is the space of all formal sums $x = -x_1 + \beta x_2 + h$, where α, β are arbitrary numbers and x_1, x_2 are abstract elements, with the usual componentwise definition of addition and multiplication by a number and with scalar product $(x, x') = \alpha\beta' + \alpha'\beta - [h, h']$,

$$x = \alpha x_1 + \beta x_2 + h, \quad x' = \alpha' x_1 + \beta' x_2 + h',$$

where $[h, h']$ is the usual scalar product in $\mathfrak{H}$.

In case IIa R is the algebra of all operators A defined by the formula $A(\alpha x_1 + \beta x_2 + h) = \lambda(A_1) \cdot \alpha \cdot x_1 + \overline{\lambda(A_1^*)} \cdot \beta \cdot x_2 + A_1 h$, $A_1 \in R_1$, and in case IIb it is the algebra of all operators A defined by the formula

$$A(\alpha x_1 + \beta x_2 + h) = \lambda\alpha x_1 + \mu\beta x_2 + A_1 h,$$

where λ, μ are arbitrary complex numbers and A_1 is any operator of R_1.

Two algebras R and $\tilde{R}$ corresponding to $\mathfrak{H}$, R_1 and $\tilde{\mathfrak{H}}$, $\tilde{R}_1$ [and additionally to $\lambda(A_1)$ and $\tilde{\lambda}(A_1)$ in case IIa] are equivalent in case IIb if and only if R_1 and $\tilde{R}_1$ are spatially isomorphic, and in case IIa if and only if there exists an isometric mapping of $\mathfrak{H}$ onto $\tilde{\mathfrak{H}}$ that transforms R_1 into $\tilde{R}_1$ and takes $\lambda(A_1)$ into $\tilde{\lambda}(A_1)$.

Case III. It has been proved that in H there is only one one-dimensional neutral invariant subspace $\mathfrak{R}$. We pick any subspace $\mathfrak{R}'$ that is piecewise-connected with $\mathfrak{R}$ and set $\mathfrak{H} = (\mathfrak{R} + \mathfrak{R}')^{\perp}$. Then $H = (\mathfrak{R} + \mathfrak{R}') \oplus \mathfrak{H}$. We pick vectors $x_0 \in \mathfrak{R}$, $x_1 \in \mathfrak{R}'$ such that $(x_0, x_1) = 1$. If $\lambda(A)$ is the eigenfunctional of R corresponding to the subspace $\mathfrak{R}$, then

$$Ax_0 = \lambda(A)\, x_0 \textit{ for all } A \in R. \tag{2.3}$$

Consider the action of R in the subspace $\mathfrak{H} + \mathfrak{R}$, which, as one readily perceives, is invariant under R. It is clear that if $h \in \mathfrak{H}$, then

$$Ah = (h, h_A)\, x_0 + A_1 h, \tag{2.4}$$

where A_1 is an operator in $\mathfrak{H}$ and h_A is a vector function on A with values in $\mathfrak{H}$. Let us put $R_1 = \{A_1 : A \in R\}$. Then R_1 turns out to be a symmetric commutative algebra in $\mathfrak{H}$ which is independent of the choice of $\mathfrak{R}'$ (up to unitary equivalence), and the correspondence $A \to A_1$ proves to be a symmetric homomorphism $R \to R_1$ which is continuous in the operator norm.

It is generally known (see [35], p. 379) that the algebra R_1 admits the following realization. The space $\mathfrak{H}$ is represented as the direct integral

$$\mathfrak{H} = \int_T \mathfrak{H}(t)\, d\sigma$$

of Hilbert spaces $\mathfrak{H}(t)$ on some Borel measure σ on the bicompact space T, and the operators $A_1 \in R_1$ act according to the formula

$$A_1 h = \{A(t)\, h(t)\},$$

where $\{h(t)\} \in \mathfrak{H}$, $A(t) \in C(t)$ and the set of all functions $A(t)$ corresponding to the operators is dense in $C(T)$.† [Recall that T is the space of maximal ideals of the closure of R_1 in the operator norm and that $A_1 \to A(t)$ is the Gel'fond isomorphism of R.]

The following subcases are possible:

IIIa. In T there exists a point t_0 such that $A(t_0) = \lambda(A)$ for all $A \in R$.

IIIb. There is no such point in T.

In case IIIa we say that the functional $\lambda(A)$ is singular, and in case IIIb we say that it is regular.

Consider case IIIa (which is the more complex). We denote by $\mathfrak{H}$ the Hilbert space $\mathfrak{H}(t_0)$ if $\sigma(\{t_0\}) \neq 0$;‡ if, however, $\sigma(\{t_0\}) = 0$, we set $\mathfrak{D} = \{0\}$. Also, we set $T_1 = T - \{t_0\}$. Then, clearly,

$$\mathfrak{H} = \int_{T_1} \mathfrak{H}(t)\, d\sigma \oplus \mathfrak{D}, \tag{2.5}$$

$$H = (\mathfrak{N} + \mathfrak{N}') \oplus \int_{T_1} \mathfrak{H}(t)\, d\sigma \oplus \mathfrak{D}. \tag{2.6}$$

†$C(T)$ denotes the space of all continuous functions on T.

‡Here, as elsewhere, we denote by $\{t_0\}$ the set consisting of the one point t_0.

By virtue of the decomposition (2.5) of $\mathfrak{H}$ the vector function h_A of Eq. (2.4) is written in the form $h_A = \{h_A(t)\} + q_A$, where $q_A \in \mathfrak{Q}$, $\{h_A(t)\} \in \int_{T_1} \mathfrak{H}(t)\, d\sigma$. It has been proved that

$$h_A(t) = (\overline{A(t)} - \overline{\lambda(A)})\, \zeta(t),\ t \in T_1,\ \zeta(t) \in \mathfrak{H}(t), \tag{2.7}$$

and $\xi(t)$ is a σ-measurable vector function. Thus, Eq. (2.4) decomposes into the two formulas

$$A\{h(t)\} = \{x_0 \int_{T_1} [A(t) - \lambda(A)]\,(h(t), \zeta(t))\, d\sigma + \{A(t)\, h(t)\}, \tag{2.8}$$

$$Aq = (q, q_A)\, x_0 + \lambda(A)\, q, \tag{2.9}$$

where $\{h(t)\} \in \int_{T_1} \mathfrak{H}(t)\, d\sigma$, $q \in \mathfrak{Q}$. It has also been proved that the set $\mathfrak{L} = \{q_A, A \in R\}$ is a linear manifold in $\mathfrak{Q}$ and that there exists an anti-isometric operator V such that $V^2 = E$ and

$$Vq_A = q_{A^*}.$$

It has been proved, finally, that

$$Ax_1 = \gamma_A x_0 + \lambda(A)\, x_1 + \{h_{A^*}(t)\} + q_{A^*}, \tag{2.10}$$

where γ_A is a numerical function on R.

Equations (2.3), (2.7), (2.9), and (2.10) completely describe the algebra R. We see that the operator $A \in R$ is determined by the system $\omega = \{A(t), q, \gamma\}$, where $A(t) \in R_1$, $q \in \mathfrak{L}$, $\gamma \in C$ (C is the field of complex numbers). We denote by $\overline{\Omega}$ the set of all such systems ω and by Ω the set of all $\omega \in \overline{\Omega}$ corresponding to operators of R. If $\omega \in \Omega$, we denote the corresponding operator $A \in R$ by A_ω. It readily follows from the correspondence $\Omega \leftrightarrow R$ that Ω has the following properties: Ω is linear (in the sense of the operations of componentwise addition and multiplication by a number); if $\omega = \{A(t), q, \gamma\} \in \Omega$, then $\omega^* = (A(t), Vq, \overline{\gamma}\} \in \Omega$; if $\omega_1 = \{A_1(t), q_1, \gamma_1\}$, $\omega_2 = \{A_2(t), q_2, \gamma_2\}$, then $\omega_3 = \{A_1(t)A_2(t), \bar{\lambda}_2 q_1 + \lambda_1 q_2,$

$$\lambda_1\gamma_2+\lambda_2\gamma_1-\int_{T_1}(A_1(t)-\lambda_1)(A_2(t)-\lambda_2)(\zeta(t),\zeta(t))\,d\sigma+(Vq_2, q_1)\}\in\Omega.$$

We refer to a set Ω with the above properties as the defining set corresponding to the decomposition (2.6) of R_1, the manifold $\mathfrak{L}$, the operator $V:\mathfrak{L}\to\mathfrak{L}$, and the vector function $\xi(t)\in\mathfrak{H}(t)$.

We have thus deduced the following:

Theorem 5. In case IIIa the algebra R is defined by the following objects: 1) the bicompact space T with isolated point t_0; 2) the measure σ with support T; 3) the symmetric algebra R_1 of functions $A(t)\in C(T)$, which is dense in $C(T)$; 4) the direct integral $\int_{T_1}\mathfrak{H}(t)\,d\sigma$ of Hilbert spaces $\mathfrak{H}(t)$, where $T_1 = T-\{t_0\}$; 5) the Hilbert space $\mathfrak{Q}$ (which can be $=0$); 6) the linear subset $\mathfrak{L}$ (which can be $=0$) with antiisometric operator $V:\mathfrak{L}\to\mathfrak{L}$ satisfying the condition $V^2=E$; 7) the σ-measurable function $\xi(t)\in\mathfrak{H}(t)$ $(t\in T_1)$, which satisfies the condition $\{A(t)-A(t_0)\zeta(t)\}\in\int_{T_1}\mathfrak{H}(t)\,d\sigma$ for any function $A(t)\in R_1$; 8) the defining set Ω corresponding to R_1, the manifold $\mathfrak{L}\subseteq\mathfrak{Q}$, the operator V, and the vector function $\xi(t)$.

Here R is defined as follows: The space H is realized in the direct sum form (2.6), and to every $\omega\in\Omega$ there corresponds an operator A_ω acting in H by Eqs. (2.3), (2.7)-(2.10).

We are now in a position to set down equivalence conditions on two algebras realized according to Theorem 5 (see [34], Theorem 3).

2.2. Commutative Symmetric Algebras of Operators in Spaces of Type Π_k $(k\geq 2)$. Let R be an algebra of the type indicated in the heading. By Corollary 1 there exists in H a nonnegative vector $x_0\neq 0$ such that

$$Ax_0=\lambda(A)\,x_0 \text{ for all } A\in R. \tag{2.11}$$

The function $A\to\lambda(A)$ is called an eigenfunctional of R and is a homomorphism of R in the field of complex numbers.

We define the root linear set belonging to the eigenfunctional $\lambda(A)$ as the set $\mathfrak{S}_\lambda = \{x: (A - \lambda(A))^p x = 0$ for all $A \in R$ and some integer $p \geq 1\}$; p can depend on x. It is easily shown that $\mathfrak{S}_\lambda \cap \mathfrak{S}_{\lambda_1} = \{0\}$, if $\lambda \neq \lambda_1$ and that $\mathfrak{S}_\lambda \perp \mathfrak{S}_{\lambda_1}$ if $\overline{\lambda(A)} \not\equiv \lambda_1(A^*)$; in particular, if $\lambda(A)$ is a nonreal functional, then $\mathfrak{S}_\lambda$ is a neutral linear set.

Let us assume at first that nonreal functionals exist (it is seen at once that there are finitely many). It has been proved that if $A \to \lambda(A)$ is one such functional, then $\overline{\lambda(A^*)}$ is also an eigenfunctional. Let $\{\lambda_j(A), \mu_j(A) \equiv \overline{\lambda_j(A^*)}, j = 1,\ldots, m\}$ be all pairwise-distinct nonreal functionals of R. It has been proved that there exist piecewise-connected pairs of subspaces $\mathfrak{R}_j$, $\mathfrak{R}_j'$ such that: a) $\mathfrak{R}_j \subseteq \mathfrak{S}_{\lambda_j}$, $\mathfrak{R}_j' \subseteq \mathfrak{S}_{\mu_j}$; b) $\mathfrak{R}_j$, $\mathfrak{R}_j'$ are invariant under R; c) the subspaces $H_j = \mathfrak{R}_j + \mathfrak{R}_j'$ are pairwise-orthogonal; d) the restriction of R to the subspace $\mathfrak{M} = (H_1 + \ldots + H_m)\perp$ does not have nonreal eigenfunctionals. Excluding the trivial case when $\mathfrak{M}$ is a Hilbert subspace, we arrive at the problem of describing the algebras that have only real eigenfunctionals.

Now let all eigenfunctionals of R be real. Let $\mathfrak{P}$ be a k-dimensional nonnegative subspace invariant under R. Then $\mathfrak{P} = \mathfrak{P}_1 + \ldots + \mathfrak{P}_p$, where $\mathfrak{P}_j$ is a root subspace of R in $\mathfrak{P}$; to every $\mathfrak{P}_j$ there corresponds an eigenfunctional $\lambda_j(A)$ such that $(A - \lambda_j(A))^{\rho_j} x = 0$ for all $x \in \mathfrak{P}$ and $A \in R$, where $\rho_j = \dim \mathfrak{P}_j$,. The number ρ_j and the functional $\lambda_j(A)$ do not depend on the choice of subspace $\mathfrak{P}$. To proceed with the description of R we introduce a special form of decomposition of H associated with R. For this we set $Q_j = \{x: x \in H, (A - \lambda_j(A))^{\rho_j} x = 0$ for all $A \in R\}$, $Q = Q_1 + \ldots + Q_p$, $\mathfrak{M} = Q^\perp$ and $\mathfrak{N} = \mathfrak{M} \cap Q$. Also, let $\mathfrak{N}'$ be a certain subspace piecewise-connected with $\mathfrak{N}$, and let $\Pi = Q \cap (\mathfrak{N}')^\perp$, $\mathfrak{H} = \mathfrak{M} \cap (\mathfrak{N}')^\perp$. It is readily shown that $\mathfrak{H}$ is a Hilbert space with scalar product $[x, y] = -(x, y)$ and that the metric in Π is not degenerate. The following decompositions hold:

$$Q = \Pi \oplus \mathfrak{N} = \Pi + \mathfrak{N}, \tag{2.12}$$

$$\mathfrak{M} = \mathfrak{H} \oplus \mathfrak{N} = \mathfrak{H} + \mathfrak{N}, \tag{2.13}$$

$$H = \Pi \oplus \mathfrak{H} \oplus (\mathfrak{N} + \mathfrak{N}'). \tag{2.14}$$

We refer to Q as the principal subspace, and to $\mathfrak{H}$ as the basic subspace of R.

Now we are ready to describe the algebra R.

Consider the restriction of R to the invariant subspace $\mathfrak{R}$. We have the decomposition $\mathfrak{R} = \mathfrak{R}_1 + \ldots + \mathfrak{R}_q$, where $\mathfrak{R}_j$ is a root subspace of R in the subspace $\mathfrak{R}$. In every $\mathfrak{R}_j$ we pick a basis $\{x_{jl}, l = 1, \ldots, r_j = \dim \mathfrak{R}_j\}$ in which the matrix of operators $A \in$ R is triangular.

$$Ax_{jl} = \sum_{s=1}^{l} \lambda_{jls} x_{js}, \quad j = 1, \ldots, q; \; l = 1, \ldots, r_j. \tag{2.15}$$

Note that $\lambda_{jll}(A) \equiv \lambda_j(A)$.

Next, consider the action of R in $\mathfrak{M}$. It is clear that

$$Ah = \sum_{j=1}^{q} \sum_{l=1}^{r_j} (h, h_{jl}) x_{jl} + A_1 h, \quad h \in \mathfrak{H}, \tag{2.16}$$

where h_{jl} is a vector function on R with values in $\mathfrak{H}$ and A_1 is an operator in $\mathfrak{H}$. We infer from the invariance of the subspaces $\mathfrak{M}$ and $\mathfrak{R}$ that the correspondence $A \to A_1$ is a homomorphism of R onto a commutative symmetric algebra R_1 in $\mathfrak{H}$, where R_1 (which we call the basic part of R) is independent of the choice of subspace $\mathfrak{R}'$, up to unitary equivalence.

Consider the action of R in the subspace Q. Using the invariance of Q and $\mathfrak{R}$ and reasoning exactly as above, we arrive at the homomorphism $A \to A_2$ of R onto a symmetric algebra R_2 in Π, which we call the principal part of R. Consequently, there exist vector functions π_{jl} such that

$$A\pi = \sum_{j=1}^{q} \sum_{l=1}^{r_j} (\pi, \pi_{jl}) x_{jl} + A_2 \pi \; \textit{for any} \; \pi \in \Pi. \tag{2.17}$$

We now explicate the structure of the algebra R_2. The decomposition $Q = Q_1 + \ldots + Q_p$ goes over under the natural mapping $\mathfrak{M} \to \mathfrak{M}/\mathfrak{R} \approx \Pi$ into the decomposition $\Pi = \Pi^1 \oplus \ldots \oplus \Pi^p$, where every Π^j is invariant under R_2 and the metric in Π^j is nondegenerate. If A_2 is a restriction of the operator $A_2 \in R_2$, then $(A_2^j - \lambda(A))^{\rho_j} = 0$. With this type of structure R_2 is said to be semidegenerate (see Sec. 2.4 concerning semidegenerate algebras). Finally we consider the action of operators $A \in \mathrm{R}$ on the vectors $y \in \mathfrak{R}'$. In $\mathfrak{R}'$ we pick a basis $\{y_{jl}\}$ biorthogonal to the basis $\{x_{jl}\}$ in $\mathfrak{R}$; this is possible, because $\mathfrak{R}$ and $\mathfrak{R}'$ are piecewise-connected. It has been proved that

$$Ay_{jl} = \sum_{m=1}^{q} \sum_{n=1}^{r_m} \bar{\alpha}_{mnjl}(A) + \sum_{n=1}^{r_j} \overline{\lambda_{jnl}}(A^*)\, y_{jn} + h_{jl}(A^*) + \pi_{jl}(A^*), \quad (2.18)$$

where $\alpha_{mnjl}(A)$ is a numerical function on R.

Equations (2.15)-(2.18) completely determine the algebra R. Consequently, every operator $A \in \mathrm{R}$ is determined by a system $\omega = \{\alpha, \lambda, A_1, A_2, \mathrm{h}, \pi\}$, where $\alpha = \{\alpha_{mnjl},\ m = 1, \ldots q;\ l = 1, \ldots r_j;\ n = 1, \ldots, r_m\}$, $\lambda = \{\lambda_{jls},\ j = 1, \ldots, q;\ l,\ s = 1, \ldots, r_j\}$, $\lambda_{jls} = 0$ for $l < s$, $\mathrm{h} = \{h_{jl},\ j, l = 1, \ldots, q\}$, $\pi = \{\pi_{jl},\ j, l = 1, \ldots, q\}$, $h_{jl} \in \mathfrak{H}$, $\pi_{jl} \in \Pi$. We denote by Ω the set of all such systems ω corresponding to all possible operators $A \in \mathrm{R}$. The symmetry and commutativity conditions on R impose a number of relations on the elements $\omega \in \Omega$ (we shall not enumerate them here; they may be found in [48], p. 164). We call any set of systems $\omega = \{\alpha, \lambda, A_1, A_2, \mathrm{h}, \pi\}$ that satisfy those relations a defining manifold associated with decomposition (2.14) and with the algebras R_1 and R_2. Every $\omega \in \Omega$ is set in correspondence by Eqs. (2.15)-(2.18) with an operator A, which we denote by A_ω. The following is obtained:

Theorem 6 [48]. Every commutative symmetric algebra R that has only real eigenfunctionals in the space Π_k is defined by the following objects: 1) a decomposition of the form (2.4) associated with the algebra; 2) the commutative symmetric algebra R_1 in $\mathfrak{H}$; 3) the semidegenerate algebra R_2 in Π; 4) the defining manifold Ω associated with decomposition (2.14) and the algebras R_1, R_2. In this case R consists of all operators $A = A_\omega$ acting according to Eqs. (2.15)-(2.18). Conversely, if we choose: 1) any decomposition of the form (2.14); 2) commutative symmetric algebra R_1 in $\mathfrak{H}$; 3) semidegenerate algebra R_2 in Π; 4) defining manifold Ω associated with decomposition (2.14) and with R_1, R_2, then the operators A_ω $(\omega \in \Omega)$ defined according to (2.15)-(2.18) form a commutative symmetric algebra in Π_k.

Note that R_1 admits the canonical realization $A \to \{A(t)\, h(t)\}$, where $h = \{h(t)\}$ comprises elements of the Hilbert space $\mathfrak{H} = \int_T \mathfrak{H}(t)\, d\sigma$,

$A(t) \in C(T)$, and T is the space of maximal ideals of the algebra $\bar{R}$ (see [36], p. 460). Let $\lambda_1, \ldots, \lambda_r$ be singular functionals of R, and let $t_1, \ldots, t_r$ be the corresponding points of T [as in Sec. 2.1, singularity of the functional λ_j means that $\lambda_j(A) \equiv A(t_j)$]. Then $h_{jl}(A) = h_{jl}(A, t)$, where $h_{jl}(A, t) \in \mathfrak{H}(t)$ for σ-almost all $t \in T$. It has been proved that

$$h_{jl}(A, t) = (\overline{A(t) - \lambda_j(A)})\,\zeta_{jl}(t) - \sum_{m=1}^{r_j} \overline{\lambda_{jml}(A)}\,\zeta_{jm}(t) \qquad (t \neq t_j), \quad (2.19)$$

where $\zeta_{jl}(t)$ is a σ-measurable vector function on $T_1 = T - \{t_1, \ldots, t_r\}$† with values in $\mathfrak{H}(t)$.

Therefore, Eqs. (2.15)-(2.18) maybe rewritten

$$Ax_{jl} = \sum_{s=1}^{r_j} \lambda_{jls}(A)\,x_{js}, \qquad (2.20)$$

$$A\{h(t)\} = -\sum_{j=1}^{q}\sum_{l=1}^{r_j}\left[\int_T (h(t),\ h_{jl}(A, t))\,d\sigma\right] x_{jl} + \{A(t)\cdot h(t)\}, \quad (2.21)$$

$$A\pi = \sum_{j=1}^{q}\sum_{l=1}^{r_j} (\pi, \pi_{jl}(A))\,x_{jl} + A_2\pi, \quad A_2 \in R_2, \qquad (2.22)$$

$$Ay_{jl} = \sum_{m=1}^{q}\sum_{n=1}^{r_m} \alpha_{jlmn}(A^*)\,x_{mn} + \sum_{n=1}^{r_j} \overline{\lambda_{jnl}(A^*)}\,y_{jn} + \{h_{jl}(A^*, t)\} + \pi_{jl}(A^*), \qquad (2.23)$$

where $h_{jl}(A, t)$ is defined by Eq. (2.19) σ-almost everywhere on $T - \{t_1, \ldots, t_r\}$.

The problem of the equivalence of two symmetric commutative algebras in the space Π_k for $k \geq 2$ has been solved by Naimark in [30]. In particular, the equivalence conditions for two canonical models have been found in that paper.

2.3. Simplification of the Canonical Model; the Case of Algebras Closed in the Norm. In [45] Langer presents a somewhat modified description of a commutative

†We denote by $\{t_1, \ldots, t_r\}$ the set consisting of the points $t_1, \ldots, t_r$.

algebra of symmetric operators in Π_k. As already remarked, it is sufficient to limit the discussion to algebras, all of whose eigenfunctionals are real. Langer also uses a decomposition of Π_k analogous to the decomposition of Naimark (see Sec. 2.2), but the corresponding subspaces are determined somewhat differently. Denoting by s(A) the spectrum of the restriction of the operator $A = A^* \in$ R onto a $\bar{k}$-dimensional nonnegative invariant subspace (it exists by a theorem of L. S. Pontryagin), Langer shows that there exists a subspace $\mathfrak{S}_R$ which is maximal invariant under R such that the spectrum of the restriction of any $A = A^* \in$ R to $\mathfrak{S}_R$ coincides with s(A). (It turns out that $\mathfrak{S}_R$, in general, is broader than the "principal subspace" Q introduced by Naimark.) Let $\mathfrak{L}_R = \mathfrak{S}_R^{\perp}$. Let us assume, as usual, that

$$\Pi_k = \mathfrak{H} \oplus \Pi \oplus (\mathfrak{N} \dotplus \mathfrak{N}'), \tag{2.24}$$

where $\mathfrak{N} = \mathfrak{L}_R \cap \mathfrak{S}_R$, $\mathfrak{N} \oplus \mathfrak{H} = \mathfrak{L}$, $\Pi \oplus \mathfrak{N} = \mathfrak{S}_R$. Under this decomposition of Π_k the algebra R is realized as an algebra of 4×4 operator matrices, each matrix A being determined by the six elements $\omega_A = (A_{23}, A_{22}, A_{11}, A_{44}, A_{13}, A_{43})$. As in Naimark's work, the set Ω_R of such "sextuples" is described (Ω_R is called a strictly-defining manifold). The virtue of Langer's description is the fact that if R is constructed on any decomposition of the form (2.24) and any associated strictly-defining manifold Ω, and then (using a theorem of Langer) a decomposition (2.24) and corresponding strictly-defining manifold Ω_R are constructed on R, it turns out that the new decomposition of Π_k is equivalent (in the natural sense) to the original decomposition, and (if these decompositions are considered to coincide) Ω and Ω_R are identical. It turns out for the model constructed by Naimark that this does not hold in general.

Naimark has noted [48] that the problem of describing the commutative symmetric algebra generated by one operator $A = A^*$ is intimately related to the notion of the spectral decomposition of A. In the cited paper of Langer substantial use is made of the results of Krein and Langer [15] bearing on that notion.

Loginov has shown [19] that the linear set $\mathfrak{S}_\lambda$ (see Sec. 2.2) is a closed subspace, and he has modified Naimark's model, adopting as the "principal subspace" (see Sec. 2.2) the subspace $\sum_\lambda + \mathfrak{S}_\lambda$ (λ runs through all eigenfunctionals of R). It turns out in this case that if the functional $\lambda(A)$ is regular, then $\mathfrak{S}_\lambda$ is a nondegenerate subspace; but if $\lambda(A)$ is a singular functional, then the point $t_0 \in T$ corresponding to $\lambda(A)$ has σ-measure 0. This simplifies the model.

Loginov has described the commutative symmetric algebras in Π_1-space which are closed under the operator norm [18]. In order to summarize this result we denote by $\mathfrak{H}$ the Hilbert space realized in the form of the direct integral of Hilbert spaces

$$\mathfrak{H} = \int_T \mathfrak{H}(t)\, d\sigma,$$

where T is a bicompact space and σ is Borel measure on T. Let $\mathfrak{H}_0$ be a two-dimensional space of type Π_1 with basis x_0, x_1, where $(x_0, x_0) = (x_1, x_1) = 0$, $(x_0, x_1) = 1$, and $H = \mathfrak{H} \oplus \mathfrak{H}_0$. In H we introduce the Π_1-metric, setting $(x, y) = -[x, y]$, where $x, y \in \mathfrak{H}$ and $[x, y]$ is the original scalar product in $\mathfrak{H}$. We define a regular algebra as an algebra of operators in H acting according to the formulas $Ax_0 = \lambda x_0$, $Ax_1 = \gamma x_0 + \lambda x_1$, $A\{h(t)\} = \{A(t)\, h(t)\}$, $\{h(t)\} \in \mathfrak{H}$, $A(t) \in C(T)$, where λ and γ are arbitrary complex numbers.

We now construct another algebra, which we call singular. Let $\mathfrak{H} = \int_{T_1} \mathfrak{H}(t)\, d\sigma \oplus Q$, where $T_1 = T - \{t_0\}$ and Q is a Hilbert space; as above, we construct the space $H = \mathfrak{H} \oplus \mathfrak{H}_0$ and introduce the Π_1-metric in it. Then a singular algebra consists of all operators acting according to the formulas

$$Ax_0 = A(t_0)\, x_0,\ Ax_1 = \gamma x_0 + A(t_0)\, x_1 + (A(t) - A(t_0))\, \zeta(t),$$

$$Aq = (q, Vl)\, x_0 + A(t_0)\, q,$$

$$A\{h(t)\} = \int_{T_1} (A(t) - A(t_0))\, (h(t), \zeta(t))\, d\sigma \cdot x_0 + \{A(t)\, h(t)\},\ h(t) \in \int_{T_1} \mathfrak{H}(t)\, d\sigma,$$

$$A(t) \in C(T),\ \int |A(t) - A(t_0)|^2\, \| \zeta(t) \|^2\, d\sigma < \infty,$$

where l runs through the subspace $\mathfrak{L} \subseteq Q$, V is an antiisometric operator in $\mathfrak{L}$, and $V^2 = E$.

Loginov has proved that any algebra closed in the norm (excluding trivial cases) is isomorphic to either a regular or a singular algebra of the type described [18].

In the case of the Π_k-metric, $k \geq 2$, it has been proved [19] that if R does not have any singular functionals, it is isomorphic to the direct sum of a complete commutative symmetric algebra of operators in a Hilbert space and a semidegenerate algebra in a space of type Π_k (see Sec. 2.4 with regard to semidegenerate algebras).

2.4. Degenerate and Semidegenerate Algebras. An algebra R is said to be degenerate if there exists a homomorphism $\lambda(A)$ of R in the field of complex numbers and a natural m such that $(A - \lambda(A))^m = 0$ for all $A \in R$. In [31] Naimark has described all (up to equivalence) commutative symmetric degenerate algebras of operators in spaces of type Π_k.

If the space $H = \Pi_k$ in which R acts decomposes into a direct orthogonal sum of invariant subspaces $H_1, \ldots, H_j$, where restrictions A_i of the operators $A \in R$ form a degenerate algebra R_i in H_i, then the algebra R is said to be semidegenerate (see Sec. 2.3).

§ 3. Group Representations Which Are Unitary in Indefinite Metric

We now assume that G is a locally-bicompact group with countable base of neighborhoods and that H is a separable space of type Π_k. We know (see, e.g., [35], p. 391) that every continuous representation of G unitary in an ordinary Hilbert space can be represented as the direct integral of irreducible unitary representations of G. If the representation is nonunitary, then such a decomposition does not hold in general; there exist reducible representations which are not decomposable into a direct sum of representations (the structure of an arbitrary nonunitary representation of a noncompact group has been meagerly investigated; the representations of "finite rank" have been described for the case of the Lorentz group [3, 4]; see also [12]). The theory of group representations in spaces of type Π_k, $1 \leq k < \infty$, may be regarded as a first approach to the study of nonunitary group representations. Although the introduction of the Π_k-metric (for $1 \leq k < \infty$) immensely simplifies the description of the representations, we still run up against a characteristic attribute of nonunitary representations, namely the occurrence of reducible but nondecomposable representations.

3.1. Representations Unitary in Π_1-Metric [25, 28]. Let $g \rightarrow U_g$ be a representation, unitary in Π_1-metric, of the group G in the space H. We denote by M the algebra of all bounded operators in H that commute with all operators U_g, $g \in G$. It is clear that M is symmetric, i.e., if $A \in M$, then $A^* \in M$, where A^* is the adjoint of the operator A under the indefinite scalar product (x, y). Let $\mathfrak{A}$ be a maximal symmetric commutative subalgebra in M. By the separability of the space there exists in $\mathfrak{A}$ a countable subset $\mathfrak{A}'$ dense in strong operator topology. Let R be the operator algebra, closed on the norm, generated by the operators of $\mathfrak{A}'$ and the unit operator 1. Then $\mathfrak{A}$ is the closure of R in strong (and weak) operator topology. According to Corollary 1, there exists in H a nonnegative one-dimensional subspace $\mathfrak{N}$ invariant under all $A \in R$. Consequently,

$$Ax_0 = \lambda(A)x_0, \quad x_0 \in \mathfrak{N}, \quad A \in R, \tag{3.1}$$

where $\lambda(A)$ is a homomorphism of R in the field of complex numbers. Inasmuch as $\mathfrak{A} = \overline{R}$ (in strong topology), Eq. (3.1) is also valid for all $A \in \mathfrak{A}$. Applying the results of § 2 to R, we infer the following as the only possible cases:

I. There exists a positive subspace $\mathfrak{N}$ invariant under all $A \in R$ (so that $\mathfrak{N}$ is also invariant under all $A \in \mathfrak{N}$).

II. There exists one and only one pair of neutral piecewise-connected one-dimensional subspaces $\mathfrak{N}$, $\mathfrak{N}'$ invariant under R, where

$$\begin{gathered} Ax_0 = \lambda(A)x_0, \quad Ax_1 = \overline{\lambda(A^*)}\,x_1, \\ x_0 \in \mathfrak{N}, \quad x_1 \in \mathfrak{N}', \quad (x_0, x_1) = 1. \end{gathered} \tag{3.2}$$

Relation (3.2) also holds for all $A \in \mathfrak{A}$.

III. There exists one and only one nonnegative subspace $\mathfrak{N}$ invariant under R; that subspace is neutral, and the corresponding eigenfunctional $\lambda(A)$ is real.

We now examine each case; we propose to use the canonical realization of R derived in § 2 for each case.

Suppose that case I occurs. We denote the corresponding functional by $\lambda(A)$ and set $H^0 = \{x : Ax = \lambda(A)x \text{ for all } A \in R\}$, $H^1 = (H^0)^{\perp}$. Then H^0 is either one-dimensional (i.e., $H^0 = \mathfrak{N}$) or of type Π_1, H^1 is a Hilbert space with scalar product $[x, y] = -(x, y)$, and $H = H^0 \oplus H^1$. It has been proved that H^0 and H^1 are invariant under all U_g, and we denote by U_g^0 and U_g^1 the restrictions of U_g to H^0 and H^1. We now analyze each of them.

We denote by $M^{(j)}$ the set of all bounded operators in H^j that commute with all operators U_g^j (j = 0, 1). It has been proved that $M^{(0)} = \{\alpha E\}$; this means that U_g^0 is an operator-irreducible representation.

Now consider the representation U_g^1. For every $A \in \mathfrak{A}$ we denote by A_1 its restriction to H^1; we say also that $\mathfrak{A}_1 = \{A_1, A \in \mathfrak{A}\}$ and $R_1 = \{A_1, A \in R)$. Then it can be proved that $\mathfrak{A}_1$ is a maximal symmetric commutative subalgebra in M_1 and R_1 is a separable closed subalgebra of $\mathfrak{A}_1$. But then also the canonical realization of R_1 corresponds to the decomposition $U_g^1 = \int_T U_g(t)\, d\sigma$ of U_g^1, where T is a bicompact space, σ is the measure on T, and $U_g(t)$ is an irreducible unitary representation of G for σ-almost all $t \in T_1$. We have thus arrived at the following:

Theorem 7. In case I the canonical realization of the algebra R corresponds to the representation $g \to U_g$ as the orthogonal sum of the representation $g \to U_g^0$, which is either one-dimensional and unitary or is operator-irreducible and unitary in Π_1-space, and the representation $g \to U_g^1$, which is the direct integral

$$U_g^1 = \int_T U_g(t)\, d\sigma$$

of representations which are continuous and unitary in the conventional sense and are irreducible for σ-almost all $t \in T$.

Consider case II. We set $H^0 = \mathfrak{N} + \mathfrak{N}'$, $H^1 = (H^0)^{\perp}$. Then $H = H^0 \oplus H^1$. The task is to prove, as above, that this decomposition of H corresponds to the decomposition of U_g into the direct orthogonal sum of U_g^0 and U_g^1. The representation U_g^0 is determined by the equations

$$U_g^0 x_0 = \tau(g)\, x_0, \quad U_g^0 x_1 = \overline{\tau(g^{-1})}\, x_1, \tag{3.3}$$

where $x_0 \in \mathfrak{N}$, $x_1 \in \mathfrak{N}'$ and $\tau(g)$ is a character (possibly nonunitary) of G. Analogous arguments lead to the following:

Theorem 8. In case II the canonical realization of the algebra R corresponds to the realization of the representation $g \to U_g$ as the orthogonal sum of the two-dimensional representation determined by (3.3) in a two-dimensional space of type Π_1 and the ordinary unitary representation $g \to U_g^1$ in a negative space which is the direct integral

$$U_g^1 = \int_T U_g(t)\, d\sigma \tag{3.4}$$

of ordinary unitary continuous representations $g \to U_g(t)$ irreducible for σ-almost all $t \in T$.

Let case III (the most complex) prevail. Then the one-dimensional space $\mathfrak{N}$ is invariant under U_g $(g \in G)$, so that

$$U_g x_0 = \tau(g)\, x_0, \quad x_0 \in \mathfrak{N}, \quad g \in G,$$

where $\tau(g)$ is a character of G. It is seen at once that $|\tau(g)| \equiv 1$. Let $\mathfrak{N}'$ be a one-dimensional null space piecewise-connected with $\mathfrak{N}$.

We shall assume that R is realized in the form of the canonical model indicated in Sec. 2.1. Let us assume also that the functional $\lambda(A)$ is singular; we refer to this as case III_a. The case (which we case III_b) when $\lambda(A)$ is regular is far simpler.

The canonical realization of R indicated in Theorem 5 leads to the decomposition $H = (\mathfrak{N} + \mathfrak{N}') \oplus \mathfrak{H}$, where $\mathfrak{N}$, $\mathfrak{N}'$ are one-dimensional piecewise-connected subspaces and $\mathfrak{N}$ is invariant under R; this implies that $\mathfrak{N}$ is also invariant under all U_g. It turns out that the subspace $\mathfrak{N} \oplus \mathfrak{H}$ is also invariant under all U_g, so that

$$U_g h = (h, h(g))\, x_0 + U_g^1 h, \quad h \in \mathfrak{H},$$

where $x_0 \in \mathfrak{N}$, $x_0 \neq 0$, and $g \to U_g^1$ is a unitary representation of G in $\mathfrak{H}$.

But, according to (2.5), $\mathfrak{H} = \int_{T_1} \mathfrak{H}(t)\, d\sigma \oplus Q$, $T_1 = T - \{t_0\}$, where the Hilbert space Q can prove to be null. It runs out that this decomposition of $\mathfrak{H}$ leads to the representation decomposition $U_g^1 =$

$\int_{T_1} U_g(t)\, d\sigma \oplus U_g(t_0)$, where $U_g(t)$ is irreducible for σ-almost all $t \in T_1$. Also, $h(g) = h_1(g) + q(g)$, where $q(g) \in Q$, and $h_1(g) \in \int_{T_1} \mathfrak{H}(t)\, d\sigma$, so that $h_1(g) = \{h(g, t)\}$, and $h(g, t) \in \mathfrak{H}(t)$ for σ-almost all $t \in T_1$. We infer from the formulas for the operators $A \in R$ in the canonical realization (see Sec. 2.1) and the commutativity condition of the operators $A \in R$ and U_g, $g \in G$, that $h(g, t) = [U_g^*(t) - \tau(g)]\, \xi(t)$, whre $\xi(t)$ is a function from the canonical realization of R (see Theorem 5).

If case III_b occurs, then $Q = 0$, and the vector x_1 may be chosen so that $\xi(t) \equiv 0$ (see Theorem 5).

Thus, we obtain:

Theorem 9. Let case III_a hold, and let

$$H = \int_{T_1} \mathfrak{H}(t)\, d\sigma \oplus Q, \quad T_1 = T - \{t_0\}, \quad \{\zeta(t)\}$$

be the Hilbert space, locally-bicompact space, and vector function, respectively, determining the canonical realization of R. Then the representation $g \to U_g$ is given by the formulas

$$U_g x_0 = \tau(g)\, x_0 \tag{3.5}$$

$$U_g\{h(t)\} = -\int ((U_g(t) - \tau(g)\, E)\, h(t),\ \zeta(t))\, d\sigma \cdot x_0 + \{U_g(t)\, h(t)\}, \tag{3.6}$$

$$U_g q = (q,\ q_g)\, x_0 + U_g(t_0)\, q, \tag{3.7}$$

$$U_g x_1 = \alpha(g^{-1})\, x_0 + \tau(g)\, x_1 + \{(U_g(t) - \tau(g)\, E)\, \zeta(t)\} + q(g^{-1}), \tag{3.8}$$

where the representations $g \to U_g(t)$ are irreducible for σ-almost all $t \in T_1$, $\tau(g)$ is a unitary character of G, $\alpha(g)$ is a numerical function on G, and $q(g)$ is a vector function on G with values from Q.

If $\sigma(\{t_0\}) = 0$, then $Q = 0$, and Eq. (3.7) and the last term in (3.8) must be discarded. If case III_b holds, then the representation

$g \to U_g$ is given by Eqs. (3.3) and (3.4), i.e., U_g is the orthogonal sum of a two-dimensional representation unitary in Π_1-metric and an ordinary unitary representation in a negative space.

3.2. Group Representations Unitary in Π_k-Metric ($k \geq 2$) [23, 26]. Let $g \to U_g$ be a representation of G unitary in a space H with Π_k-metric, where $k \geq 2$. The algebras M, $\mathfrak{A}$, and R are defined as in Sec. 3.1. The following proposition is easily proved:

Every root linear set of R is invariant under all U_g ($g \in G$).

We assume in the beginning that there exist nonreal functionals of the algebra R; let $\{\lambda_j(A), \mu_j(A), j = 1,\ldots, m\}$ be all distinct nonreal eigenfunctionals, where $\lambda_j(A) \equiv \overline{\mu_j(A^*)}$, let $\mathfrak{R}_j, \mathfrak{R}_j'$ be the corresponding root subspace, and let $\overline{H} = \sum_{j=1}^{m} (\mathfrak{R}_j \dotplus \mathfrak{R}_j')$ be a hyperbolic subspace of R. Then $H = \overline{H} \oplus \overline{H}^{\perp}$, and this decomposition leads to the decomposition of U_g into two representations U_g^0 and U_g^1 acting in $\overline{H}$ and $\overline{H}^{\perp}$, respectively. It follows from the fact that $\mathfrak{R}_j$ and $\mathfrak{R}_j'$ are mutually adjoint. We obtain the following:

Theorem 10. The restriction of the representation U_j to a hyperbolic subspace of the algebra R decomposes into the direct orthogonal sum of a finite number of representations, each of which is the direct sum of two mutually-adjoint representations in piecewise-connected spaces.

It now remains to examine the most complicated situation, when all the eigenfunctionals $\lambda_j(A)$, $j = 1,\ldots, m$, of R are real. Let Q, $\mathfrak{R}$, $\mathfrak{R}'$, $\mathfrak{M}$, $\mathfrak{H}$, Π be the subspaces defined in H for the algebra in Sec. 2.2. Then

$$\mathfrak{M} = \mathfrak{R} \oplus \mathfrak{H}, \tag{3.9}$$

$$Q = \mathfrak{R} \oplus \Pi, \tag{3.10}$$

$$H = (\mathfrak{R} + \mathfrak{R}') \oplus \mathfrak{H} \oplus \Pi. \tag{3.11}$$

Using the decomposition (3.11) of H, we realize R in the form of the canonical model indicated in Sec. 2.2. We wish to find out how the operators of U_g are specified under this realization of R.

Consider the restriction of U_g to the subspace $\mathfrak{R}$. Let $\mathfrak{R}_j$, j = 1,..., q be root subspaces of R in $\mathfrak{R}$; then $\mathfrak{R} = \mathfrak{R}_1 + \ldots + \mathfrak{R}_q$, and every $\mathfrak{R}_j$ is invariant under all U_g. We choose a certain basis $\{x_{jl}, j = 1,\ldots, q, l = 1,\ldots, r_j\}$ in $\mathfrak{R}_j$. Then the restrictions of U_g to $\mathfrak{R}$ are given by the formulas

$$U_g x_{jl} = \sum_{s=1}^{r_j} v_{jls}(g)\, x_{js}, \tag{3.12}$$

where $v_{jls}(g) = (U_g x_{jl}, y_{js})$, $\{y_{js}\}$ is the basis in $\mathfrak{R}'_j$ biorthogonal to $\{x_{jl}\}$. Next we consider the restriction of U_g to the subspace $\mathfrak{M}$. It follows from decomposition (3.9) that

$$U_g h = \sum_{j=1}^{q} \sum_{l=1}^{r_j} (h, h_{jl}(g))\, x_{jl} + U_g^1 h, \quad h \in \mathfrak{H}, \tag{3.13}$$

where $h_{jl}(g)$ is a continuous vector function on G with values in $\mathfrak{H}$ and U_g^1 is an operator in $\mathfrak{H}$. Since the subspaces $\mathfrak{M}$ and $\mathfrak{R}$ are invariant under the representation, the correspondence $g \to U_g^1$ is a unitary representation of G in $\mathfrak{H}$; we refer to it as the principal part of the representation U_g. It can be proved that the representation $g \to U_g^1$ is independent, up to unitary equivalence, of the choice of subspace $\mathfrak{R}'$ piecewise-connected with $\mathfrak{R}$.

We now recall the decomposition $\mathfrak{H} = \int_T \mathfrak{H}(t)\, d\sigma$, indicated in Sec. 2.2, where T is a locally bicompact space with measure σ. It has been proved that this decomposition of the space $\mathfrak{H}$ leads to the decomposition of U_g^1 into a direct integral $U_g^1 = \int_T U_g(t)\, d\sigma$, where the correspondence $g \to U_g(t)$ is a continuous unitary representation of G for σ-almost all $t \in T$, and it is irreducible for σ-almost all $t \in T_1 = T - \{t_1,\ldots, t_r\}$.

Setting $h = \{h(t)\}$, $h_{jl}(g, t) = \{h_{jl}(g, t)\}$, where $h(t) \in \mathfrak{H}(t)$ for σ-almost all $t \in T$, we can rewrite Eq. (3.13) as follows:

$$U_g \{h(t)\} = \sum_{j=1}^{q} \sum_{l=1}^{r_j} \left[\int_T (h(t), h_{jl}(g, t))\, d\sigma \right] x_{jl} + \{U_g(t)\, h(t)\}. \tag{3.14}$$

From the formulas for the operators $A \in R$ in the canonical model and the condition $AU_g = U_g A$ for all $A \in R$ and $g \in G$ we find that

$$h_{jl}(g, t) = U^*_g(t)\zeta_{jl}(t) - \sum_{m=1}^{r_j} \overline{v_{jml}(g)}\zeta_{jm}(t), \tag{3.15}$$

for σ-almost all $t \in T - \{t_1,\ldots, t_p\}$, where $\zeta_{jl}(t)$ is a function from the canonical model of R (see Theorem 6).

Now consider the restriction of U_g to the space Q. As above, we infer from the invariance of the subspaces Q and $\mathfrak{R}$ under all U_g that

$$U_g\pi = \sum_{j=1}^{q}\sum_{l=1}^{r_j}(\pi, \pi_{jl}(g))\, x_{jl} + U^2_g\pi, \quad \pi \in \Pi, \tag{3.16}$$

where $\pi_{jl}(g)$ are continuous vector functions on G with values in Π and U^2_g is a continuous unitary representation of G in Π.

Consider the action of the operator U_g on the vectors $y \in \mathfrak{R}'$. For this it suffices to state $U_g y_{jl}$. We can prove that

$$U_g y_{jl} = n_{jl}(g^{-1}) + \sum_{n=1}^{r_j}\overline{v_{jml}(g^{-1})}\, y_{jn} + \{h_{jl}(g^{-1}, t)\} + \pi_{jl}(g^{-1}), \tag{3.17}$$

where $n_{jl}(g)$ is a continuous vector function on G with values in $\mathfrak{R}$.

Assembling the results thus far, we arrive at the following:

<u>Theorem 11.</u> Let $g \to U_g$ be a continuous unitary representation of the group G in a separable space H of type Π_k, $2 \le k < \infty$, M the set of all bounded operators that commute with all operators U_g, $\mathfrak{A}$ a maximal commutative symmetric subalgebra in H, and R a symmetric subalgebra in $\mathfrak{A}$ which is closed and separable in the operator norm, contains the unit operator, and is dense in $\mathfrak{A}$ in weak operator topology. Let all eigenfunctionals of R be real.

Then for the canonical realization of R according to Eqs. (2.20)-(2.23) of Sec. 2.2 the operators U_g are specified by Eqs. (3.12)-(3.16), where: a) $g \to \|v_{jls}(g)\|$ for every $j = 1,\ldots, q$ is a continuous representation of G; b) $g \to U^2_g$ is a continuous unitary representation of G in the space Π; c) $g \to U_g(t)$ is a continuous unitary representation of G irreducible for σ-almost all $t \in T - \{t_1, \ldots, t_r\}$, and $U_g(t)$ is a measurable operator function of t on T for every $g \in G$; d) $n_{jl}(g)$ and $\pi_{jl}(g)$ are continuous vector functions on G with values in $\mathfrak{N}$ and Π, respectively; e) the vector function $h_{jl}(g, t)$ in Eq. (3.14) is defined for σ-almost all $t \in T - \{t_1, \ldots, t_r\}$ by Eq. (3.15).

3.3. Representations of Symmetric Algebras in Spaces of Type Π_k. The symmetric representations of symmetric algebras can be investigated by the same scheme as the unitary representations of groups. Let M be the algebra of all bounded operators in H that commute with all operators $U_A (a \in A)$, $\mathfrak{A}$ a maximal commutative symmetric subalgebra in M, and R a separable and closed (in operator norm topology) symmetric subalgebra which is dense in weak operator topology and contains the unit operator. Applying the reasoning of Sec. 3.2 to the representations $a \to U_A$, we arrive at the following proposition:

Theorem 12. The assertions of Theorem 10 remain valid when the representation $g \to U_g$ of a group G is replaced by the representation $a \to U_A$ of the algebra A.

Thus, by "stripping off" the hyperbolic subspace we need only confine ourselves to the case when all eigenfunctionals of the algebra R are real. The arguments of Sec. 2.2 lead to the following result:

Theorem 13. The assertions of Theorem 11 remain valid when the continuous representations $g \to U_g$ of a group G are replaced by continuous symmetric representations of a separable symmetric normed algebra A.

3.4. Structure of Symmetric Operator in Π_k-Space. Let A be an algebra of bounded linear operators, separable on the operator norm, in a separable space H. Let M be the set of all bounded operators that commute with all $a \in A$, $\mathfrak{A}$ a maximal symmetric commutative subalgebra in M, and R a symmetric subalgebra in $\mathfrak{A}$ which is separable and closed on the operator norm, contains the unit operator, and is dense in $\mathfrak{A}$ in weak operator topology. Applying Theorem 13 to the identity representation $a \to a$, we obtain the realization of A corresponding to the canonical realization of R. We shall omit the specific formulas for the operators $A \in R$ in that realization (see [26], Theorem 5).

3.5. Representations of Commutative Groups and Algebras; Representations of Solvable Groups. From the general theorems of Secs. 3.1 and 3.2 we obtain a description of the representations of commutative locally-bicompact groups and commutative algebras with involution in spaces of type Π_k (see [23]).

Note that in this case we have emerging as the space T (see Theorems 11 and 13) the group of characters of G (where a representation of the group G is involved) or the space of symmetric maximal ideals of A (where a representation of the algebra A is involved).

For connected solvable groups the following theorem holds (see [21]). Let $g \rightarrow U_g$ be a unitary representation of a connected solvable group in a space Π_k, $1 \leq k < \infty$. Then: 1) there exists a nonnegative vector $x_0 \neq 0$ which is a common eigenvector for all u_g $(g \in G)$; 2) there exists a k-dimensional nonnegative subspace invariant under all U_g.

In some cases this result is also true for unconnected solvable groups, for example the group of matrices of the form $\begin{pmatrix} k & h \\ 0 & k^{-1} \end{pmatrix}$, where k and h are elements of any discrete field P (see [10]).

3.6. Representations of the Lorentz Group. Let G be the Lorentz group. In this case the description of the representations may be refined. We begin with the recollection that all (up to Naimark equivalence) completely irreducible representations in a Banach space have been found for G (see [36], p. 199). From these representations we readily sort out those which are unitary in indefinite metric (note that for the representations unitary in Π_k-metric of an arbitrary group, where $1 \leq k < \infty$, irreducibility implies complete irreducibility [7]).

Consider arbitrary (reducible) representations of G unitary in Π_k-metric. The following theorem, which has been proved independently by Naimark [32] and Ismagilov [9], shows that the nonunitary representations discretely involved in the representation $g \rightarrow U_g$ can be stripped off [an irreducible representation $g \rightarrow U_g^0$ of G is said to be discretely involved in the representation $g \rightarrow U_g$ if the space B of $g \rightarrow U_g$ contains a subspace $B_1 \subseteq B$ invariant under all U_g $(g \in G)$ such that the restriction of U, to B_1 is equivalent to U_g^0].

Theorem 14. Let $g \to U_g$ be a unitary representation of G in a space $H = \Pi_k$, and let it discretely involve a nonunitary representation $g \to U_g^0$ acting in an invariant subspace $H_1 \subseteq H$. Then:

1) the representation U_g also discretely involves the adjoint representation acting in an invariant subspace $H_2 \subseteq H$ piecewise connected with H_1;

2) there exists a subspace $H_3 = \Pi_{k'} \subset H$ invariant under all U_g and invested with the following properties:

a) The restriction of U_g onto $H \ominus H_3$ does not discretely involve nonunitary irreducible representations:

b) The restriction of U_g to H_3 is the orthogonal sum of a finite number (possibly one) representations, each of which is the direct sum of two mutually-adjoint representations acting in two piecewise-connected subspaces.

Thus, it suffices to limit ourselves to representations that do not discretely involve discretely-nonunitary representations.

The further description of the representations of G is contained in Ismagilov's paper, in which it is shown that if the trivial cases are discarded there exists for any unitary representation of G in Π_k a k-dimensional neutral invariant subspace $\mathfrak{R}$. If we set $\mathfrak{M} = \mathfrak{R}^\perp$, then $\mathfrak{M} = \mathfrak{R} \oplus \mathfrak{H}$, where $\mathfrak{H}$ is a Hilbert subspace, and the restriction $U_g^{\mathfrak{M}}$ of U_g to $\mathfrak{M}$ turns out to be the "coupling" of a finite-dimensional representation $g \to S_g$ acting in $\mathfrak{R}$ and a unitary (in the usual sense) representation $g \to V_g$ in the subspace $\mathfrak{H}$. It turns out that U_g is determined up to unitary equivalence by its restriction $U_g^{\mathfrak{M}}$ and that the decomposition of the representation $U_g^{\mathfrak{M}}$ into a finite direct orthogonal sum of representations can be extended to a decomposition of U_g. Therefore, the problem reduces to the description of the representation $U_g^{\mathfrak{M}}$.

It has been proved that $U_g^{\mathfrak{M}}$ decomposes into a finite orthogonal sum of representations, each of which is one of the following types: a) finite-dimensional; b) unitary in the usual sense; c) a nondecomposable coupling of a finite-dimensional representation which is a multiple of an irreducible nonunity representation $g \to S_g$ of G with a unitary representation which is a multiple of a unitary irreducible representation $g \to T_g$, where T_g is related to the representation S_g (see [1], p. 212); d) or, finally, a coupling of the unity representation with a unitary representation $g \to V_g$, where the decomposition of V_g into irreducible representations involves only representations of the additional series and the fundamental-series representation which is associated with the unity representation.

Now if $g \to U_g$ is a unitary representation of G in the space Π_k and U_g contains a k-dimensional neutral subspace $\mathfrak{R}$ invariant under all U_g, we can first decompose the corresponding representation (where $\mathfrak{M} = \mathfrak{R}^{\perp}$, $U_g^{\mathfrak{M}}$ is the restriction of the representation to $\mathfrak{M}$), as described above, and then extend that decomposition to all of Π_k. We thereby obtain a decomposition of the representation $g \to U_g$ into representations of a fully-defined form.

Schlieder [51] has investigated (with a view toward applications in physics) the representations of the inhomogeneous Lorentz group. He describes a class of reducible but undecomposable representations which are unitary in indefinite metric (without assuming that the indefiniteness rank k is finite). He also ascertains a certain class of reducible but not completely-reducible representations of that group which are unitary in Π_k-metric. We note that the indefinite metric in the state space of a physical system was first studied by Dirac; see also the book [47].

3.7. Representations of the Group SL(2, P), where P Is an Infinite Field. Let P be an infinite field (with discrete topology), and $g \to U_g$ an irreducible representation of the group SL (2, P) = G which is unitary in the space Π_k $(1 \le k < \infty)$. These representations have been described in a paper by Ismagilov [10]. We give the fundamental result.

It has been proved that the restrictions of a representation U_g onto a subgroup K of matrices of the form

$$k_{\alpha,\beta} = \begin{pmatrix} \alpha^{-1} & \beta \\ 0 & \alpha \end{pmatrix}, \quad \alpha \neq 0,$$

contain a one-dimensional neutral invariant subspace $\mathfrak{R}$. If $x_0 \in \mathfrak{R}$, $x_0 \neq 0$, then $U(k_{\alpha,\beta}) = \tau(\alpha)x_0$, where $\tau(\alpha)$ is a character (nonunitary) of the multiplicative group of the field P, with $\tau(\alpha) > 0$ for all $\alpha \neq 0$. It turns out that for $k \ge 2$ there exists an isomorphism of P onto a dense subfield of the field R or C (i.e., the field of real or field of complex numbers) such that U_g is continuous in the ordinary topology of R (or C) and, hence, the representation $g \rightarrow U_g$ extends to a continuous representation unitary in Π_k-metric of SL (2, R) [or SL (2, C)].

If, on the other hand, k = 1, then the function $\sqrt{\tau(x)}$ is a nontrivial norm of P, where if the norm is Archimedian, the representation once again extends to SL (2, R) or to SL (2, C). But if the norm $\sqrt{\tau(x)}$ is non-Archimedian, then U_g extends by continuity to the group SL (2, P_τ), where P_τ is the completion of P on the norm $\sqrt{\tau(x)}$. The converse also turns out to be true: Any nontrivial non-Archimedian norm $\tau_1(x)$ on P leads to an irreducible representation of SL (2, P) unitary in Π_1-metric.

Now let P be a field with a nontrivial non-Archimedian norm $\phi(x)$, and let the representation U_g of SL (2, P) = G be continuous in the appropriate topology. If $\tau(x)$ is the corresponding character of P, then $\tau(x) \equiv |\varphi(x)|^\lambda$, $\lambda > 0$; conversely, to any $\lambda > 0$ there corresponds by the rule described above an irreducible representation of G unitary in Π_1-metric. Note that for the case of a locally-compact field these results are easily obtained from the general results of the book [1] (p. 224). The case of a field P which is not locally compact has been studied by Ismagilov [11]. He proved that if P_0 is a ring of integral elements of P, i.e., if $P_0 = x: \phi(x) \leq 1$, then in the space of any irreducible representation unitary in Π_1-metric there exists a vector $h_0 \neq 0$ (defined up to a multiplier) invariant under the operators U_g, where $g \in SL(2, P_0)$. Moreover, he calculates the elementary spherical function $(U_g h_0, h_0)$. In [11] Ismagilov also considers the unitary (in the usual sense) representations of SL (2, P) when the field P is not locally compact.

4. Dissipative Operators and the Extension of Dual Subspaces (see [41, 42])

Let H be a Hilbert space with scalar product [x, y] and indefinite scalar product (x, y) = [Jx, y], $J = P_+ - P_-$. We denote by $\mathfrak{M}_+$ (or $\mathfrak{M}_-$) the set of all maximal nonnegative (or nonpositive) subspaces.

An ordered pair of subspaces (N, P) is called a dual pair of subspaces if N is nonnegative, P is nonpositive, and $N \perp P$ in the sense of the indefinite metric. If, in addition, $N \in \mathfrak{M}_+$, $P \in \mathfrak{M}_-$, then the pair (N, P) is said to be a maximal dual pair of subspaces.

Phillips has investigated the following problem: Let R be an algebra of bounded operators in H, and (N_0, P_0) a dual pair of subspaces, where N_0 and P_0 are invariant under operators of R. Does there exist a maximal dual pair of subspaces (N, P) invariant under R and such that $N \supseteq N_0$ and $P \supseteq P_0$? The commutativity and symmetry of R (in the sense of the indefinite metric) are natural restrictions.

Below we shall give some applications of this problem; for now we cite results pertinent to the problem. We denote by A^0 and A^* operators which are adjoint to A in the sense of the definite or indefinite metric, respectively. Phillips has shown [41] that the problem stated above is solved in the affirmative (for any dual pair of subspaces invariant under R) under the following conditions:

1) R is closed with respect to both involutions.

Also, Phillips has proved that the next three conditions are equivalent, each guaranteeing an affirmative solution to the problem.

2) H can be renormalized so that $A^0 = A^*$ for any $A \in R$.

3) The group G of unitary (in Π_k-metric) operators of R is bounded in the norm.

4) There exists an invariant dual pair of subspaces (N, P) such that $N \oplus P = H$.

It follows from a result of Naimark (see Theorem 2) that for $k < \infty$ the problem is solved in the affirmative (as above, the algebra R is assumed to be commutative and symmetric in Π_k-metric).

In a paper by Larionov [17] condition 3) is replaced by the following milder condition: $\| U^k \| < C_U$ $(k = \pm 1, \pm 2, \ldots)$ for any $U \in G$ (C_U depends on U).

The problem just described is applicable to the theory of the extensions of dissipative operators. Recall that an operator A defined in a Hilbert space H_0 is called a dissipative operator if $(A_0 x, x) + (x, A_0 x) \leq 0$ for all $x \in D(A_0)$.

If $G(A_0)$ is the graph of A_0 in $H = H_0 \oplus H_0$, then $G(A_0)$ is a nonpositive subspace in the sense of the Π_∞-metric $(y, z) = (y_1, z_2) + (y_2, z_1)$, where $y = \{y_1, y_2\}$ and $z = \{z_1, z_2\}$.

Let B_0 be a dissipative operator such that $A_0 \subseteq B_0^*$, i.e., $(A_0 x, y) = (x, B_0 y)$ for $x \in D(A_0)$ and $y \in D(B_0)$. The problem is to extend the pair A_0, B_0 to a pair A, B of maximal (i.e., not admitting dissipative extensions) of dissipative operators such that $A^* = B$. It follows from the condition $A_0 \subseteq B_0^*$ that $G(A_0)$ and $G(-B_0)$ form a dual pair of subspaces in H, and the problem reduces to the extension of this pair to a maximal dual pair of subspaces. The latter problem has been solved by Phillips [42], who thus obtained a complete solution to the problem of the extension of dissipative operators.

Up to now the extension problem has not involved the algebra R (it may be assumed that $R = \{\lambda E\}$, where λ is a scalar). One of the problems (indicated by Phillips) that reduce to the extension problem for invariant dual pairs of subspaces is concerned with the analysis of the extensions of a pair of dissipative operators A_0, B_0 (see above) which commute with elements of some algebra R (for example, A_0, B_0 could be formally-adjoint minimal differentiable operators, and R could be generated by a group of motions G which leave A_0 and B_0 invariant); it is required that the extensions $A \supset A_0$ and $B \supset B_0$ also commute with R.

The Phillips problem finds another application in the theory of first-order systems of partial differential equations [41, 42]. Let the differential operator

$$L_0 y = \sum_{i=1}^{n} A^i \frac{\partial}{\partial x_i} y + By$$

be given in the space of smooth vector functions $y = (y_1(x), \ldots, y_n(x))$ which are defined in a domain $\Delta \subset R_n$ and vanish in the neighborhood of the boundary of Δ; A^i, B are matrix functions in Δ, where L_0 is dissipative, so that

$$D = B + B^* + \sum_{i=1}^{n} \frac{\partial}{\partial x_i} A^i \leqslant 0.$$

The general results of Phillips imply the existence of maximal dissipative extensions $L \supset L_0$, but the most interesting problem is the description of the extensions characterized by "local boundary conditions"; this means that for any smooth function $\gamma(x)$ in Δ the condition $\gamma(x)\, y(x) \in D(L)$ holds for any $y(x) \in D(L)$. If we introduce into $H = L_2(\Delta) \oplus L_2(\Delta)$ the operators

$$\gamma(y^1, y^2) = \left(\gamma_0 y^1, \gamma_0 y_2 + \sum_{i=1}^{m} \left(\frac{\partial}{\partial x_i} \gamma_0 \right) A^i y^1 \right),$$

where γ_0 (x) is a smooth function in Δ, then the above condition is equivalent to invariance of the graph G (L) of the operator L with respect to the resulting algebra R of operators γ. We arrive at the Phillips problem.

Note that only partial results are available with regard to the Phillips problem; in [41] Phillips indicates a number of unsolved problems which are significant in connection with applications to differential equations.

§ 5. Some Unsolved Problems

1. Operator-Irreducible Representations. A representation $g \to U_g$ of a group G in a Banach space B is said to be operator-irreducible if any bounded linear operator A in B commuting with all U_g is a multiple of the unit operator. The problem is to describe the operator-irreducible representations which are unitary in Π_k-metric. For the Lorentz group certain representations of the type indicated are given in [9]. A more specific problem is posed in the latter: Give a representation $g \to V_g$ of a group G in the Hilbert space $H = H_0 \oplus H_1$, where dim $H_0 = k$, H_0 is invariant under V_g, and

$$V_g x = V_g^1 x + L_g x \quad for \quad x \in H_1, \tag{5.1}$$

where V_g^1 is a unitary representation in H_1 and L_g acts from H_1 into H_0. It is clear that U_g preserves the degenerate bilinear form (x, y) defined by the equations

$$(x, y) = -[x, y], \quad if \quad x, y \in H_1,$$
$$(x, y) = 0, \quad if \quad x \in H_0 \quad or \quad y \in H_0.$$

Let H_2 be a k-dimensional linear space. In the space $\bar{H} = H + H_2$ we introduce the indefinite scalar product (x, y) coinciding on H with the bilinear form defined above, such that H_1 and H_2 are piecewise-connected. Can U_g be extended to a representation in $\bar{H}$ unitary in this method? How are all such extensions described? For what groups can all operator-irreducible representations be obtained by this method?

We note the ultimately simple case of this problem when the restriction of the representation U_g to H_0 is a multiple of the unit operator. It can be shown that the stated problem has a solution for any coupling of the form (5.1) if $H^2(G, c) = 0$, where $H^2(G, c)$ is the second group of irreducible cohomologies of the group G with coefficients in the field C (G acts trivially in C). It has been shown for certain classes of groups that this condition is also necessary in order for our problem to be solvable for any representation of the form (5.1). (The foregoing results have been obtained by R. S. Ismagilov, but have not been published.)

2. The description given by Naimark for the representations (see Sec. 4.6) can sometimes be simplified (see the case of the Lorentz group considered above). For other groups how are the simplest representations described from which it is possible to obtain any representation as a finite direct orthogonal sum? Note that the simplest representations need not be operator-irreducible.

3. Find the unitary (in the sense of Π_k-metric) equivalence conditions for two representations unitary in Π_k-metric.

4. The representations unitary in Π_∞-metric have not been studied. Is it true that the irreducibility of a representation unitary in Π_∞-metric is equivalent to its complete irreducibility? Note that for the simplest groups (R and Z) this problem is exceedingly complex and has not been solved; in the case of the Π_k-metric $k < \infty$, the answer is affirmative (see [7]).

5. If two unitary (in the usual sense) representations are Naimark-equivalent, they are unitarily equivalent. Is this true in the case of the Π_k-metric? More precisely, let U_g^1 and U_g^2 be two representations unitary in Π_k-metric, and let H^1 and H^2 be the spaces of the representations; let U_g^1 be Naimark-equivalent to U_g^2. Does there exist a mapping V, isometric in Π_k-metric and two-way continuous, from H^1 to H^2 such that $U_g^1 V = V U_g^2$?

6. For which, not necessarily connected, solvable groups does the Naimark theorem (of Sec. 3.5) remain valid?

7. Describe all representations, unitary in Π_k-metric ($1 \leq k < \infty$), of nilpotent groups.

Literature Cited

1. Gel'fand, I. M., Graev, M. I. and Vilenkin, N. I., Integral Geometry and Related Aspects of the Theory of Representations, Fizmatgiz, Moscow (1962), 656 pages.
2. Gel'fand, I. M. and Pyatetskii-Shapiro, I. I., Theory of Representations and Automorphic Functions, Nauka, Moscow (1966), 512 pages.
3. Gel'fand, I. M. and Ponomarev, V. Ya., Nondecomposable representations of the Lorentz group, Usp. Mat. Nauk, 23(2): 3-60 (1968).
4. Zhelobenko, D. P., Linear representations of the Lorentz group, Dok. Akad. Nauk SSSR, 126(5): 935-933 (1959).
5. Iokhvidov, I. S. and Krein, M. G., Spectral theory of operators in spaces with indefinite metric (I), Trudy Moskov. Mat. Obshch., 5: 367-432 (1956).
6. Iokhvidov, I. S. and Krein, M. G., Spectral theory of operators in spaces with indefinite metric (II), Trudy Moskov. Mat. Obshch., 8: 413-396 (1969).
7. Ismagilov, R. S., Rings of operators in a space with indefinite metric, Dokl. Akad. Nauk SSSR, 171(2): 269-271 (1966).
8. Ismagilov, R. S., Description of the unitary representations of the Lorentz group in a space with indefinite metric, Dokl. Akad. Nauk SSSR, 158(2): 268-270 (1964).
9. Ismagilov, R. S., Unitary representations of the Lorentz group in a space with indefinite metric, Izv. Akad. Nauk SSSR, Ser. Mat., 30(3): 497-522 (1966).
10. Ismagilov, R. S., Irreducible representations of the discrete group SL(2, P) which are unitary in the indefinite metric, Izv. Akad. Nauk SSSR, Ser. Mat., 30(4): 923-950 (1966).
11. Ismagilov, R. S., Elementary spherical functions on the group SL(2, P) over a field P which is not locally compact, with respect to a subgroup of metrices with integral elements, Izv. Akad. Nauk SSSR, Ser. Mat., 31(2): 361-390 (1967).
12. Ismagilov, R. S., On the linear representations of the group SL(2, R), Mat. Sb., 74(4): 496-515 (1967).
13. Krein, M. G., An application of the fixed-point principle in the theory of linear transformations of spaces with indefinite metric, Usp. Mat. Nauk, 5: 180-190 (1950).
14. Krein, M. G., A new application of the fixed-point principle in the theory of operators in a space with indefinite metric, Dokl. Akad. Nauk SSSR, 154(5): 1023-1026 (1964).
15. Krein, M. G. and Langer, G. K., On the spectral function of a self-adjoint operator in a space with indefinite metric, Dokl. Akad. Nauk SSSR, 152(1): 39-42 (1963).
16. Langer, G. K., On invariant subspaces of linear operators acting in a space with indefinite metric, Dokl. Akad. Nauk SSSR, 169(1): 12-15 (1966).
17. Larionov, E. A., A commutative family of operators in a space with indefinite metric, Mat. Zametki, 1(5): 589-594 (1967).
18. Loginov, A. I., Banach commutative symmetric algebras of operators in the Pontryagin space Π_1, Dokl. Akad. Nauk SSSR, 179(6): 1276-1278 (1968).
19. Loginov, A. I., Commutative symmetric algebras of operators in a Pontryagin space, Izv. Akad. Nauk SSSR, Ser. Mat., 33(3) (1969).

20. Naimark, M. A., On commutative unitary operators in the space Π_χ, Dokl. Akad. Nauk SSSR, 149(6):1261-1263 (1963).

21. Naimark, M. A., Unitary representations of solvable groups in spaces with indefinite metric, Izv. Akad. Nauk SSSR, Ser. Mat., 27(5):1181-1185 (1963).

22. Naimark, M. A., On commutative algebras of operators in the space Π_1, Dokl. Akad. Nauk SSSR, 156(4):734-737 (1964).

23. Naimark, M. A., On the representations of commutative symmetric Banach algebras and commutative topological groups in the space Π_k, Dokl. Akad. Nauk SSSR, 170(2):271-274 (1966).

24. Naimark, M. A., On commutative algebras of operators in the space Π_k, Dokl. Akad. Nauk SSSR, 161(4):767-770 (1965).

25. Naimark, M. A., On the structure of the unitary representations of locally-bicompact groups in the space Π_1, Izv. Akad. Nauk SSSR, Ser. Mat., 29(3):689-770 (1965).

26. Naimark, M. A., On the structure of the unitary representations of locally-bicompact groups and symmetric representations of algebras in the Pontryagin spaces Π_k, Izv. Akad. Nauk SSSR, Ser. Mat., 30(5):1111-1132 (1966).

27. Naimark, M. A., Analog of Stone's theorem in a space with indefinite metric, Dokl. Akad. Nauk SSSR, 170(6):1259-1261 (1966).

28. Naimark, M. A., On the unitary representations of locally-bicompact groups in the space Π_1, Dokl. Akad. Nauk SSSR, 160(2):281-283 (1965).

29. Naimark, M. A., On the unitary equivalence conditions for commutative symmetric algebras in the space Π_k, Dokl. Akad. Nauk SSSR, 160(6):1257-1260 (1965).

30. Naimark, M. A., Unitary equivalence conditions for commutative symmetric algebras in the Pontryagin space Π_k, Trudy Moskov. Mat. Obshch., 15:383-399 (1966).

31. Naimark, M. A., Degenerate operator algebras in the Pontryagin space Π_k, Izv. Akad. Nauk SSSR, Ser. Mat., 30(6):1229-1256 (1966).

32. Naimark, M. A., On the unitary representations of the Lorentz group in spaces with indefinite metric, Mat. Sb., 65(2):198-211 (1964).

33. Naimark, M. A., On the unitary representations of the Lorentz group in the space Π_k, Dokl. Akad. Nauk SSSR, 152(5):1064-1067 (1963).

34. Naimark, M. A., Commutative operator algebras in the space Π_1, Rev. Roumaine Math. Pures Appl., 9(6):499-528 (1964).

35. Naimark, M. A., Normed Rings, Gostekhizdat, Moscow (1956), 487 pages [English edition: P. Noordhoff, Groningen (1964)].

36. Naimark, M. A., Linear Representations of the Lorentz Group, Fizmatgiz, Moscow (1958), 376 pages [English edition: Pergamon, New York (1964)].

37. Pontryagin, L. S., Hermitian operators in a space with indefinite metric, Izv. Akad. Nauk SSSR, Ser. Mat., 8(1):243-280 (1944).

38. Sobolev, S. L., On the motion of a symmetric top with a liqud-filled cavity, Zh. Prikl. Mekh. Tekh. Fiz., No. 3, pp. 20-55 (1960).

39. Fan, K., Invariant subspaces of certain linear operators, Bull. Amer. Math. Soc., 69:773-777 (1963).

40. Fan, K., A generalization of Tychonoff's fixed-point theorem, Math. Ann., 142:305-310 (1960/61).

41. Phillips, R. S., The extension of dual subspaces invariant under an algebra, Proc. Internat. Sympos. Linear Spaces, Jerusalem, 1960, Jerusalem Academic Press, Jerusalem; Pergamon, New York (1961), pp. 366-398.

42. Phillips, R. S., Dissipative operators and hyperbolic systems of partial differential equations, Trans. Amer. Math. Soc., 90:193-254 (1959).

43. Fan, Ky, Invariant cross sections and invariant linear subspaces, Israel J. Math., 2(1):19-26 (1964).

44. Fan, Ky, Invariant subspaces for a semigroup of linear operators, Proc. Konikl. Nederland. Akad. Wet., A68(3):447-451 (1965).

45. Langer, H., On a theorem of M. A. Neumark [Naimark] [in German], Math. Ann., 175(4):303-314 (1968).

46. Langer, H., On the spectral theory of I-self-adjoint operators [in German], Math. Ann., 146(1):60-85 (1962).

47. Nagy, K. L., State Vector Spaces with Indefinite Metric in Quantum Field Theory, Akad. Kiadó, Budapest (1966), 131 pages with illus.

48. Naimark, M. A., Commutative symmetric operator algebras in Pontryagin spaces Π_k [in German], Math. Ann., 162(1):147-171 (1965).

49. Naimark, M. A., On commuting unitary operators in spaces with indefinite metric, Acta Sci. Math., 24(3-4):177-189 (1963).

50. Naimark, M. A., On unitary group representations in spaces with indefinite metric, Acta Sci. Math., 26(3-4):201-209 (1965).

51. Schlieder, S., Indefinite metric in a state space and its probabilistic interpretation (I, II) [in German], Z. Naturforsch., 15a:448-467 (1960).